Phytoremediation of Domestic Wastewater with the Internet of Things and Machine Learning Techniques

Phytoremediation of Domestic Wastewater with the Internet of Things and Machine Learning Techniques highlights the most recent advances in phytoremediation of wastewater using the latest technologies. It discusses practical applications and experiences utilizing phytoremediation methods for environmental sustainability and the remediation of wastewater. It also examines the various interrelated disciplines relating to phytoremediation technologies and plots industry's best practices to share this technology widely, as well as the latest findings and strategies. It serves as a nexus between artificial intelligence, environmental sustainability and bioremediation for advanced students and practising professionals in the field.

Phytoremediation of Domestic Wastewater with the Internet of Things and Machine Learning Techniques

Hauwa Mohammed Mustafa and Gasim Hayder

CRC Press

Taylor & Francis Group

Boca Raton London New York

CRC Press is an imprint of the
Taylor & Francis Group, an **informa** business

Designed cover image: Photograph by Gasim Hayder.

MATLAB® is a trademark of The MathWorks, Inc. and is used with permission. The MathWorks does not warrant the accuracy of the text or exercises in this book. This book's use or discussion of MATLAB® software or related products does not constitute endorsement or sponsorship by The MathWorks of a particular pedagogical approach or particular use of the MATLAB® software.

First edition published 2023
by CRC Press
6000 Broken Sound Parkway NW, Suite 300, Boca Raton, FL 33487-2742

and by CRC Press
4 Park Square, Milton Park, Abingdon, Oxon, OX14 4RN

CRC Press is an imprint of Taylor & Francis Group, LLC

© 2023 Taylor & Francis Group, LLC

Library of Congress Cataloging-in-Publication Data
Names: Mustafa, Hauwa Mohammed, author. | Hayder, Gasim, author.
Title: Phytoremediation of domestic wastewater with the internet of things and machine learning techniques / Hauwa Mohammed Mustafa and Gasim Hayder.
Description: First edition. | Boca Raton : CRC Press, 2023. | Includes bibliographical references and index.
Identifiers: LCCN 2022042621 (print) | LCCN 2022042622 (ebook) | ISBN 9781032417448 (hbk) | ISBN 9781032417523 (pbk) | ISBN 9781003359586 (ebk)
Subjects: LCSH: Sewage–Purification–Data processing. | Phytoremediation–Data processing. | Internet of things. | Machine learning.
Classification: LCC TD755 .M87 2023 (print) | LCC TD755 (ebook) | DDC 628.40285–dc23/eng/20221129
LC record available at https://lccn.loc.gov/2022042621
LC ebook record available at https://lccn.loc.gov/2022042622

ISBN: 978-1-032-41744-8 (hbk)
ISBN: 978-1-032-41752-3 (pbk)
ISBN: 978-1-003-35958-6 (ebk)

DOI: 10.1201/9781003359586

Typeset in Times
by codeMantra

Contents

About the Authors

Hauwa Mohammed Mustafa is an Academic Lecturer in the Department of Pure and Applied Chemistry, Kaduna State University (KASU), Nigeria. She obtained her Doctorate Degree from Universiti Tenaga Nasional (UNITEN), Malaysia. Her research areas focus on Wastewater Treatment, Bioenergy, IoT, Artificial Intelligence, and Material Science.

Gasim Hayder is a Senior Lecturer and the Head of the Water & Environmental Engineering Unit at the Civil Engineering Department, College of Engineering, Universiti Tenaga Nasional (UNITEN), Malaysia. He is also a Chartered Engineer-MIET (CEng) from the Engineering Council (UK). He received his Ph.D. in Civil Engineering from Universiti Teknologi PETRONAS (UTP), Malaysia. His research focus is in the field of Environmental Engineering and Water and Wastewater Assessment, Treatment, Modelling, and Monitoring.

1 Impacts of Sustainable Development Goals (SDGs) in Wastewater Management

1.1 INTRODUCTION

Wastewater can be described as water that has been contaminated by anthropogenic activities. Wastewater is water that has been "used" for various purposes, including commercial, industrial, agricultural activities, domestic, storm water or sewer infiltration. The composition of wastewater varies according to the source. Similarly, wastewater is made up of biological, chemical and physical compounds that may have a detrimental effect on the water quality, rendering it unfit for human consumption and hazardous to aquatic life. Anthropogenic activities from agricultural, industrial and domestic activities release waste product such as ammonia (NH_3), phosphorous (P), nitrate (NO_3) and potassium (K) into natural water bodies. Consequently, this brings about eutrophication in the aquatic environment (Sukačová et al., 2015). Thus, complete recovery or removal of nutrients from wastewater should be treated as a fundamental and crucial process before disposal into natural bodies to protect natural water sources (Hu et al., 2012).

Furthermore, one of the key components of the 2030 sustainable development goals (SDGs) is to improve water quality. The conventional wastewater treatment methods which involve multiple treatment processes fail in the complete removal of water pollutants. These residues contain certain amount of nutrients (nitrogen, phosphorus and heavy metals) and pathogens that cause the spread of diseases such as typhoid, cholera, and skin irritation and promote the rapid growth of aquatic weed plants in rivers and lakes (Kutty et al., 2009; Meena et al., 2019). Therefore, the search and exploitation of affordable and renewable methods of wastewater treatment such as phytoremediation has become crucial. The applications of aquatic plant (macrophytes) in phytoremediation of wastewater are not new (Akinbile et al., 2015). Phytoremediation technique uses the plant root system in absorbing excess nutrients from wastewater (Nwachukwu & Osuji, 2007; Udeh et al., 2013). However, the potentials of aquatic plants in phytoremediation processes depend on the characteristics and origin of the wastewater (Ng et al., 2017). The diagram of wastewater is presented in Figure 1.1.

DOI: 10.1201/9781003359586-1

FIGURE 1.1   Wastewater.

1.2   SOURCES OF WASTEWATER IN OUR ENVIRONMENT

Wastewater is classified as industrial, municipal, agricultural and domestic waste-water. Households generate wastewater from showers, dishwashers, laundry and flush toilets. Figure 1.2 depicts the major types of wastewater (Edokpayi et al., 2017). Additionally, wastewater can be transported through combined sewer or sanitary sewer. In other words, the phrases "water reclamation" and "wastewater reuse" apply when the treated waste is used for another purpose. Furthermore, water reclamation refers to the treatment of wastewater to make it beneficial within the acceptable water quality standards. In general, water reuse is the application of reclaimed water for useful purpose. Additionally, the word "reclaimed" is frequently interchanged with "recycled water" (Asano et al., 2007).

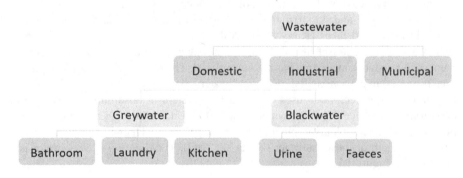

FIGURE 1.2   Classification of wastewater.

1.2.1 DOMESTIC WASTEWATER

Wastewater from households has been classified into different colours based on their usage (Shi et al., 2018). Wastewater from kitchen and bathroom usage is classified under grey water. Urinal waste with minimal flushing water is classified as yellow water, while faecal waste from the toilet is categorized as black water (De Gisi et al., 2015). Grey water is often described as urban wastewater generated from showers, hand basins and kitchen sinks. Additionally, kitchen grey water contains high BOD, variable thermotolerant coliforms' loads and cleaning agents (detergents). In contrast, bathroom grey water has a high concentration of suspended particles, hair and turbidity from cleaning chemicals such as toothpaste, soap and shampoo. Furthermore, it has lower BOD concentrations and thermotolerant coliforms loads. According to a previous study, laundry and kitchen grey water contain much more organic and physical contaminants than bathroom grey water. All grey water types have a high degree of biodegradability with regard to COD (Metcalf & Eddy, 2003). However, phosphorus and nitrogen are deficient due to the absence of faeces and urine in bathroom grey water. Similarly, laundry grey water is also deficient in phosphorus due to the application of phosphorous-free detergent in laundry. Powdered laundry detergent is the main source of pollution in grey water. Many of them have a high concentration of salts and phosphorus, and they are alkaline in nature. Long-term application of laundry water in gardens may result in stunting of plants and salt build-up in the soil (De Gisi et al., 2015).

1.2.2 INDUSTRIAL WASTEWATER

The word "industrial wastewater" refers to wastewater discharged from various industrial processes (Ahmed et al., 2021). Water is used in industries for different purposes, including manufacturing of goods, heating and cooling of industrial processes, firefighting, dissolving of solvents, and transportation of waste matter and raw materials. Pollutants such as zinc (Zn), chromium (Cr), copper (Cu), lead (Pb), iron (Fe), cadmium (Cd), nickel (Ni), arsenic (As) and mercury (Hg) can be present in industrial wastewater depending on the type of manufacturing or commercial activities carried (Wang & Yang, 2016). Furthermore, industrial wastewater is divided into inorganic and organic wastewater. In general, treating industrial wastewater is tough because it necessitates an individual analysis of the set-ups and the construction of specific treatment plants on an industry-specific basis. For example, organic contaminants are maintained in the secondary treated wastewater even after the primary biological treatment has taken place. Power plants, food/beverage industries, steel/iron production plants, industrial laundries, oil and gas fracking plants, metal finishers and mining are examples of industries that generate wastewater (Liu et al., 2014).

1.2.3 MUNICIPAL WASTEWATER

Municipal wastewater is a mixture of domestic and industrial wastewater (Guldhe et al., 2017). The characteristics of municipal wastewater depend on the source, and it is significantly influenced by local activities. Municipal wastewater contains organic compounds and nutrients such as phosphorus, nitrogen and basic metal ions essential for aquatic plant growth. In general, municipal wastewater contains less phosphorus and nitrogen than agricultural wastewater. Animal wastewater is commonly nutrient-dense, particularly in nitrogen and phosphorus (Zhu et al., 2019). Additionally, the concentration of nutrients in agricultural wastewater remains highly variable depending on the wastewater source. Furthermore, agricultural wastewater contains fungicides, insecticides, herbicides and antibiotics (Cai et al., 2013).

1.3   STRATEGIES FOR EFFECTIVE MANAGEMENT OF WASTEWATER

The management of wastewater is an essential part of comprehensive integrated water resources management. The majority of water users rely on sufficient levels of water quality. Water quality protection and management are increasingly becoming the focus of major efforts and costs in water management (Loucks & van Beek, 2005). Consequently, wastewater management has a long history. Generally, the difficulties encountered in wastewater management are largely due to the absence of uniform definitions, reporting guidelines and a central repository for wastewater treatment data (Malik et al., 2015). Therefore, wastewater management entails proper handling of wastewater in order to protect the environment and ensure economic, public health and social stability (Metcalf & Eddy, 2003). Water protection and sanitization are the main goals of wastewater management. This means that water must be sufficiently clean to be used for drinking and washing, irrigation and commercial purposes by industries. Nevertheless, different strategies are required for efficient wastewater management depending on the location, source of the wastewater, population size, technical capacity, economic development and system of governance in place. Additionally, the strategies can differ based on the level of quality needed for safe disposal and end users. Thus, wastewater management should reflect the economic, social and environmental needs of the end users (Corcoran et al., 2010). Additionally, economic assessment and evaluation must be discussed in wastewater management research with the objective of connecting natural resource with the population and benefits to the wider community (Ćetković et al., 2022). Moreover, effective approaches of wastewater management would address global water scarcity challenges and relieve pressure on natural water bodies. Notwithstanding, implementation of wastewater management must take into account different wastewater treatment methods, stakeholders, environmental laws and policies of waste disposal, as well as any external costs and benefits associated with the reuse choice (Garcia & Pargament, 2015).

1.4   CONTRIBUTIONS OF SUSTAINABLE DEVELOPMENT GOALS (SDGs) IN PROMOTING WASTEWATER MANAGEMENT

Ever since the adoption of the sustainable development goals (SDGs), the crucial problems listed have gained worldwide recognition. The United Nations (UN) considered the SDGs for water due to its significance to humans and aquatic species.

The UN adopted the 2030 Agenda for sustainable development and its 17 goals in September 2015, which served as a great driving force to global efforts in achieving sustainable development (UN, 2018). In other words, the SDGs are universal, aspirational, communicable and measurable targets that put countries on a path to embrace and achieve sustainable growth between 2015 and 2030 (Malik et al., 2015). In addition, SDG 6 aims "to improve water quality by reducing pollution, eliminating dumping and minimizing the release of hazardous chemicals and materials and improving wastewater treatment and increasing recycling and safe reuse globally" (UN, 2018). Particularly, SDG 6.3 is concerned with wastewater treatment. Hence, wastewater remediation is one of the ways in which SDG 6 hopes to provide clean water and sanitation. As such, efforts are being made around the world to ensure proper wastewater treatment methods. Additionally, the problem with the release of untreated wastewater into bodies of water is that it causes the extinction of aquatic life, which is also an important goal of the SDG, to preserve aquatic lives. Thus, wastewater treatment is of great importance.

However, through the work of the UN in light of the SDGs, in the year 2021, reports were collected from around the world on the wastewater situation of different countries. According to the indicators, billions of people worldwide lack access to clean water and adequate sanitation, 10% continue to live in regions with severe water shortages and untreated sewage is released into the environment. These indicators have only become available as a result of the SDG's impact. Additionally, through the SDGs, adequate awareness and attention is being brought to the wastewater crisis, as with awareness comes change. Furthermore, through awareness created by SDGs, a lot of countries and non-governmental organizations around the world are providing services geared towards the mitigation of wastewater.

Moreover, another multifaceted problem of wastewater management is further aggravated by climate change. Climate change is also one of the concerns of the SDGs (SDG 13 aims to combat climate change and its impacts) (UN, 2018). Consequently, climate change has a dual effect on wastewater management and water resources through changes in precipitation, temperature, storm-related changes and sea level rise. Hence, extreme weather events caused by climate change will result in more untreated wastewater. As a result, the need for wastewater management is increasingly mandatory (Singh & Tiwari, 2019). Therefore, SDG of combating climate change would positively improve wastewater management.

1.5 CONCLUSION

The recycling of wastewater is spreading throughout the world, particularly in areas where water shortage restricts the growth of economic sectors like manufacturing and agriculture. In addition, water reuse prevents the discharge of treated wastewater into water bodies, as it offers an alternative to other investments made to preserve natural water bodies. Nevertheless, different strategies are required for efficient wastewater management depending on the location, source of the wastewater, population size, technical capacity, economic development and system of governance in place. Additionally, strategies for wastewater management can differ based on the level of quality needed for safe disposal and end users. Thus, wastewater management should reflect the economic, social and environmental needs of the end users. Furthermore, the successful achievement of the SDGs would contribute immensely in sustainable wastewater management, thereby ensuring clean and adequate water supply, and safe environment for all.

REFERENCES

Ahmed, J., Thakur, A., & Goyal, A. (2021). Industrial Wastewater and Its Toxic Effects. In *Biological Treatment of Industrial Wastewater*. Series: Chemistry in the Environment (pp. 1–14). https://doi.org/10.1039/9781839165399-00001

Akinbile, C. O., Ogunrinde, T. A., Che bt Man, H., & Aziz, H. A. (2015). Phytoremediation of domestic wastewaters in free water surface constructed wetlands using Azolla pinnata. *International Journal of Phytoremediation*, *18*(1), 54–61. https://doi.org/10.1080/1522 6514.2015.1058330

Asano, T., Burton, F. L., Leverenz, H. L., & Tsuchihashi, R., Tchobanoglous, G. (2007). *Water Reuse: Issues, Technologies, and Applications*. McGraw Hill, New York.

Cai, T., Park, S.., & Li, Y. (2013). Nutrient recovery from wastewater streams by microalgae: Status and prospects. *Renewable and Sustainable Energy Reviews*, *19*, 360–369. https://doi.org/10.1016/j.rser.2012.11.030

Ćetković, J., Kneževic, M., Lackic, S., Zarkovic, M., Vujadinovic, R., Zivkovic, A., & Cvijovic, J. (2022). Financial and economic investment evaluation of wastewater treatment plant. *Water (Switzerland)*, *14*(122), 1–23. https://doi.org/10.3390/w14010122

Corcoran, E., Nellemann, C., Baker, E., Bos, R., Osborn, D., & Savelli, H. (2010). Sick water? The central role of waste-water management in sustainable development. A rapid response assessment. Mine Water and the Environment, *30*(3), 169–174. https://doi.org/10.1007/s10230-011-0140-x

De Gisi, S., Casella, P., Notarnicola, M., & Farina, R. (2015). Grey water in buildings: A mini-review of guidelines, technologies and case studies. *Civil Engineering and Environmental Systems*, *33*(1), 1–20. https://doi.org/10.1080/10286608.2015.1124868

Edokpayi, J. N., Odiyo, J. O., & Durowoju, O. S. (2017). Impact of Wastewater on Surface Water Quality in Developing Countries: A Case Study of South Africa. In Hlanganani Tutu (Ed), *Water Quality* (pp. 401–416). IntechOpen, London, United Kingdom. https://doi.org/10.5772/66561

Garcia, X., & Pargament, D. (2015). Resources, conservation and recycling reusing wastewater to cope with water scarcity: Economic, social and environmental considerations for decision-making. *Resources, Conservation & Recycling, 101*, 154–166. https://doi.org/10.1016/j.resconrec.2015.05.015

Guldhe, A., Kumari, S., Ramanna, L., Ramsundar, P., Singh, P., Rawat, I., & Bux, F. (2017). Prospects, recent advancements and challenges of different wastewater streams for microalgal cultivation. *Journal of Environmental Management*, *203*, 299–315. https://doi.org/10.1016/j.jenvman.2017.08.012

Hu, Z., Houweling, D., & Dold, P. (2012). Biological nutrient removal in municipal wastewater treatment: New directions in sustainability. *Journal of Environmental Engineering*, *March*, 307–317. https://doi.org/10.1061/(ASCE)EE.1943-7870.0000462.

Kutty, S. R. M., Ngatenah, S. N. I., Mohamed, H. I., & Malakahmad, A. (2009). Nutrients removal from municipal wastewater treatment plant effluent using Eichhornia crassipes. *World Academy of Science, Engineering and Technology, 60*, 1115–1123.

Liu, S., Ma, Q., Wang, B., Wang, J., & Zhang, Y. (2014). Advanced treatment of refractory organic pollutants in petrochemical industrial wastewater by bioactive enhanced ponds and wetland system. *Ecotoxicology*, *23*(4), 689–698. https://doi.org/10.1007/s10646-014-1215-9

Loucks, D. P., & van Beek, E. (2005). Water Quality Modeling and Prediction. In *Water Resource Systems Planning and Management*. https://doi.org/10.1007/978-3-319-44234-1_10

Malik, O. A., Hsu, A., Johnson, L. A., & De Sherbinin, A. (2015). A global indicator of wastewater treatment to inform the Sustainable Development Goals (SDGs). *Environmental Science and Policy*, *48*, 172–185. https://doi.org/10.1016/j.envsci.2015.01.005

Meena, R. A. A., Yukesh, K. R., Sindhu, J., Ragavi, J., Kumar, G., Gunasekaran, M., & Rajesh Banu, J. (2019). Trends and resource recovery in biological wastewater treatment system. *Bioresource Technology Reports*, *7*(March), 1–16. https://doi.org/10.1016/j.biteb.2019.100235

Metcalf & Eddy. (2003). *Wastewater Engineering: Treatment and Reuse* (4th ed.). McGraw Hill, New York.

Ng, Y. S., Samsudin, N. I. S., & Chan, D. J. C. (2017). Phytoremediation capabilities of Spirodela polyrhiza and Salvinia molesta in fish farm wastewater: A preliminary study. *IOP Conference Series: Materials Science and Engineering*, *206*(1). https://doi.org/10.1088/1757-899X/206/1/012084

Nwachukwu, E. O., & Osuji, J. O. (2007). Bioremedial degradation of some herbicides by indigenous white rot fungus, Lentinus subnudus. *Journal of Plant Sciences*, *2*(6), 619–624. https://doi.org/10.3923/jps.2007.619.624

Shi, K.-W., Wang, C.-W., & Jiang, S. C. (2018). Quantitative microbial risk assessment of Greywater on-site reuse. *Science of the Total Environment*, *635*, 1507–1519. https://doi.org/10.1016/j.scitotenv.2018.04.197

Singh, S., & Tiwari, S. (2019). Climate Change, Water and Wastewater Treatment: Interrelationship and Consequences. In R. Singh, A. Kolok, & S. Bartelt-Hunt (Eds.), *Water Conservation, Recycling and Reuse: Issues and Challenges*. Springer, Singapore. https://doi.org/10.1007/978-981-13-3179-4_11

Sukačová, K., Trtílek, M., & Rataj, T. (2015). Phosphorus removal using a microalgal biofilm in a new biofilm photobioreactor for tertiary wastewater treatment. *Water Research*, *71*, 55–63. https://doi.org/10.1016/j.watres.2014.12.049

Udeh, N. U., Nwaogazie, I. L., & Momoh, Y. (2013). Bio-remediation of a crude oil contaminated soil using water hyacinth (Eichhornia crassipes). *Advances in Applied Science Research*, *4*(2), 362–369. http://www.pelagiaresearchlibrary.com/advances-in-applied-science/vol4-iss2/AASR-2013-4-2-362-369.pdf

United Nations. (2018). *The 2030 Agenda and the Sustainable Development Goals an opportunity for Latin America and the Caribbean (LC/G.2681-P/Rev.3)*, Santiago. https://repositorio.cepal.org/bitstream/handle/11362/40156/25/S1801140_en.pdf

Wang, Q., & Yang, Z. (2016). Industrial water pollution, water environment treatment, and health risks in China. *Environmental Pollution*, *218*, 358–365. https://doi.org/10.1016/j.envpol.2016.07.011

Zhu, S., Huo, S., & Feng, P. (2019). Developing Designer Microalgal Consortia: A Suitable Approach to Sustainable Wastewater Treatment. In M. Alam & Z. Wang (Eds.), *Microalgae Biotechnology for Development of Biofuel and Wastewater Treatment* (pp. 569–598). Springer Nature, Singapore. https://doi.org/10.1007/978-981-13-2264-8_22

2 Solving the Shortage of Clean Water through Wastewater Treatment

2.1 INTRODUCTION

Wastewater treatment is primarily concerned with the protection of public health and environment. In the past, there was no specific method for wastewater treatment. As an alternative, wastewater from buildings was channelled into canals and gutters, where it was finally transferred into bodies of water such as streams, rivers, lakes and sea (Cisneros et al., 2018). This natural treatment method based on dilution was sufficient due to lower population density and industrial activities, which resulted in the release of lesser pollution load into the environment compared to present times. Presently, localities that rely on polluted water for drinking and other domestic use frequently suffer eutrophication and disease epidemics (Jasim, 2020). Additionally, it has been shown that wastewater contains significant concentrations of a range of organic contaminants, necessitating the adoption of extra treatment stages and control mechanisms (Armah et al., 2020). Furthermore, wastewater treatment plants (WWTPs) are designed to ensure that the final effluent is free from pollutants such as biochemical oxygen demand (BOD), total suspended solids (TSS) and chemical oxygen demand (COD) (Jasim, 2020). Advances in research have led to the introduction of new approaches in removing excess nutrients from wastewater (Al-Ghouti et al., 2019). Among them, the conventional treatment techniques are characterized by high energy consumption, long processes, carbon emission, odour, high amount of sludge discharge and high cost. These act as barriers to the idea of sustainable growth in wastewater treatment (Li et al., 2019; Sun et al., 2016). On the contrary, biological methods feature ecologically engineered natural treatment systems due to their cost-effectiveness, process flexibility (energy-efficient/solar-driven) and environmentally friendly advantages compared to other treatment methods (Ma et al., 2016; Zhu et al., 2008). The view of a wastewater treatment plant (WWTP) is shown in Figure 2.1.

2.2 CONVENTIONAL SEWAGE TREATMENT TECHNOLOGY

Conventional wastewater treatment involves the removal of coliform bacteria, suspended solids, biochemical oxygen demand (BOD), and nutrients such as phosphate-phosphorous, nitrate-nitrogen, nitrite-nitrogen and ammonia-nitrogen from wastewater. BOD can be removed through the oxidation of organic matter to water

DOI: 10.1201/9781003359586-2

FIGURE 2.1   View of wastewater treatment plant (WWTP).

and CO_2 by microorganisms (Jasim, 2020). At the same time, suspended solids are primarily eliminated through physical sedimentation. Similarly, the removal of dissolved phosphorus and nitrogen is essential in nutrient-removal wastewater treatment systems. The toxicity of non-ionized ammonia compounds to fish and other aquatic species is another consequence of nitrogen compounds in wastewater effluents. However, conventional wastewater treatment is divided into preliminary, primary, secondary and tertiary stage (Asiwal et al., 2016). The brief description of preliminary, primary, secondary and tertiary stages is presented in the next section.

2.3   PRELIMINARY TREATMENT OF WASTEWATER

In the preliminary phase of wastewater treatment, solid debris transported by sewers are removed. These solid debris could be floating particles such as wood, faeces, grains of rock or sand and rags. The sewage is passed through holes or chambers to remove floating debris, and at regular intervals, the built-up waste on the bars is scraped off (Jasim, 2020). Grits are removed by reducing the flow velocity to a point where the silt and grit settle, while the organic solids remain suspended (Abdel-Raouf et al., 2019).

2.4 PRIMARY TREATMENT OF WASTEWATER

In the primary treatment stage, the wastewater is transferred into sedimentation tanks to remove the settleable solids by gravity. A well-built sedimentation tank can be used in the removal of BOD. Pathogen removal during primary treatment varies depending on the removal rates of different organisms (Singh et al., 2021).

2.5 SECONDARY TREATMENT OF WASTEWATER

Secondary treatment of wastewater tends to reduce the BOD concentration by removing organic materials. This is accomplished mostly through a heterotrophic bacterial population that uses constituents for energy and growth. The aerobic oxidation of BOD is accomplished through a variety of biological unit processes. These biological processes are categorized based on whether their microbial population is growing in a fixed film or in a dispersed film. Biofilms are bonded to a fixed surface in the fixed biofilm reactors and the organic molecules are absorbed and broken down aerobically. Whereas in dispersed film, microorganism growth reactors are allowed to mix freely with the wastewater and are retained in suspension by mechanical agitation during the treatment process. Numerous researchers have demonstrated that biological systems can eliminate over 90% of pollutants present in wastewater (Asiwal et al., 2016).

2.6 TERTIARY TREATMENT OF WASTEWATER

The tertiary treatment method was designed to completely eliminate all organic compounds through chemical or biological methods (Jasim, 2020). The biological approach functions better than chemical processes, which are generally too expensive to install in most locations and may result in secondary contamination. Modern treatment techniques are mainly complex procedures such as carbon adsorption, chemical precipitation, reverse osmosis or ozonation. These approaches include those that target specific nutrients, such as phosphorus or nitrogen (Asiwal et al., 2016). Therefore, primary and secondary treatment methods are used in WWTPs to remove organic waste and easily settled contaminants in wastewater. However, due to the discharge of heavy metals and refractory organics, the secondary treated effluent is laden with phosphorus and inorganic nitrogen contributing to eutrophication and other long-term concerns (Abdel-Raouf et al., 2012). The stages of wastewater treatment are presented in Figure 2.2 (Adapted from Asiwal et al., 2016).

2.7 BIOLOGICAL METHOD OF WASTEWATER TREATMENT

Over the last 50 years, different methods for abatement of pollution have been introduced. Additionally, since the last two decades, researchers have been concerned about the possibility of using macrophytes in the biological treatment of wastewater (Moore et al., 2009). These approaches, which favour biological-based treatment over traditional procedures, are supported by government regulations and popular opinion (Ekperusi et al., 2019). Anaerobic treatment, Anammox technology, algal

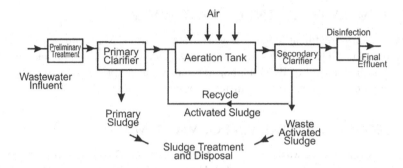

FIGURE 2.2   Stages of wastewater treatment.

technology and microbial fuel cells (MFCs) are some of the biological technologies that have been used in the treatment of wastewater. On the other hand, these methods have drawbacks such as complex mode of operation, time consumption, high cost and energy demand. According to McCarty et al. (2011), high energy of about 0.3–0.6 kWh/m^3 was consumed in the treatment of wastewater using activated sludge method. Additionally, Singh et al. (2016) reported that 0.78 kg CO_2 eq/m^3 was the average carbon intensity for WWTPs. As a result, high energy requirement poses a major setback on the net economic burden in applications of biological methods in WWTPs (Li et al., 2017). Therefore, development of sustainable and energy-neutral technologies for applications in WWTPs has become crucial. Hence, different countries around the world are employing cheap and eco-friendly technologies such as phytoremediation methods in wastewater treatment (George & Gabriel, 2017). Phytoremediation technique of wastewater treatment using hydroponic systems is cost-effective, energy efficient and simple to operate (Worku et al., 2018).

2.8 CONCLUSION

Globally, improving water infrastructure is necessary, since water conservation is crucial to securing a reliable supply of clean water. Water is an essential resource not only for the survival of all species, but also for human economic growth. Large volume of wastewater is produced from agriculture, households and industries. Wastewater reuse has been identified as an attractive solution to the global problem of water shortage. Wastewater can be reused after treatment for different beneficial purposes. Wastewater that has been properly treated for drinkable purpose provides a new source of water. Furthermore, adoption of wastewater remediation would ease global water shortages while promoting clean water supply and protecting the environment.

REFERENCES

Abdel-Raouf, M. E., Maysour, N. E., Farag, R. K., & Abdul-Raheim, A.-R. M. (2019). Wastewater treatment methodologies, review article. *International Journal of Environment & Agricultural Science, 3*(1), 1–25.

Abdel-Raouf, N., Al-Homaidan, A. A., & Ibraheem, I. B. M. (2012). Microalgae and wastewater treatment. *Saudi Journal of Biological Sciences*, *19*(3), 257–275. https://doi.org/10.1016/j.sjbs.2012.04.005

Al-Ghouti, M. A., Al-Kaabi, M. A., Ashfaq, M. Y., & Da'na, D. A. (2019). Journal of water process engineering produced water characteristics, treatment and reuse : A review. *Journal of Water Process Engineering*, *28*(January), 222–239. https://doi.org/10.1016/j.jwpe.2019.02.001

Armah, E. K., Chetty, M., Adedeji, J. A., Kukwa, D., Mutsvene, B., Shabangu, K. P., & Bakare, B. F. (2020). Emerging Trends in Wastewater Treatment Technologies: The Current Perspective. In I. A. Moujdin, & J. K. Summers (Eds.), *Promising Techniques for Wastewater Treatment and Water Quality Assessment*. IntechOpen, London, United Kingdom, 1–28. https://doi.org/10.5772/intechopen.93898.

Asiwal, R. S., Sar, S. K., Singh, S., & Sahu, M. (2016). Wastewater treatment by effluent treatment plants. *SSRG International Journal of Civil Engineering (SSRG – IJCE)*, *3*(12), 29–35.

Cisneros, L., Ibanescu, M., Keen, C., Lobato-Calleros, O., & Niebla-Zatarain, J. (2018). Bibliometric study of family business succession between 1939 and 2017: Mapping and analyzing authors' networks. Scientometrics, 117(2). https://doi.org/10.1007/s11192-018-2889-1

Ekperusi, A. O., Sikoki, F. D., & Nwachukwu, E. O. (2019). Application of common duckweed (Lemna minor) in phytoremediation of chemicals in the environment: State and future perspective. *Chemosphere*, *223*, 285–309. https://doi.org/10.1016/j.chemosphere.2019.02.025

George, G. T., & Gabriel, J. J. (2017). Phytoremediation of heavy metals from municipal waste water by Salvinia molesta Mitchell. *The Saudi Journal of Life Sciences*, *2*(3), 108–115. https://doi.org/10.21276/haya

Jasim, N. A. (2020). The design for wastewater treatment plant (WWTP) with GPS X modelling. *Cogent Engineering*, *7*(1). https://doi.org/10.1080/23311916.2020.1723782

Li, K., Liu, Q., Fang, F., Luo, R., Lu, Q., Zhou, W., Huo, S., Cheng, P., Liu, J., Addy, M., Chen, P., Chen, D., & Ruan, R. (2019). Microalgae-based wastewater treatment for nutrients recovery : A review. *Bioresource Technology*, *291*(June). https://doi.org/10.1016/j.biortech.2019.121934

Li, W., Li, L., & Qiu, G. (2017). Energy consumption and economic cost of typical wastewater treatment systems in Shenzhen, China. *Journal of Cleaner Production*, *163*, S374–S378. https://doi.org/10.1016/j.jclepro.2015.12.109

Ma, B., Wang, S., Cao, S., Miao, Y., Jia, F., Du, R., & Peng, Y. (2016). Biological nitrogen removal from sewage via anammox: Recent advances. *Bioresource Technology*, *200*, 981–990. https://doi.org/10.1016/j.biortech.2015.10.074

McCarty, P. L., Bae, J., & Kim, J. (2011). Domestic wastewater treatment as a net energy producer-can this be achieved? *Environmental Science and Technology*, *45*(17), 7100–7106. https://doi.org/10.1021/es2014264

Moore, G. T., Hertzler, P., Dufresne, L., Clifford, R., Barnard, J., David, S., & Brown, J. (2009). *Nutrient Control Design Manual, State of Technology Review Report*. Cadmus Group Inc.; USEPA, Watertown, MA.

Singh, P., Kansal, A., & Carliell-Marquet, C. (2016). Energy and carbon footprints of sewage treatment methods. *Journal of Environmental Management*, *165*, 22–30. https://doi.org/10.1016/j.jenvman.2015.09.017

Singh, S., Sharma, A., & Malviya, R. (2021). Industrial wastewater: Health concern and treatment strategies abstract . *The Open Biology Journal*, *9*(2), 1–10. https://doi.org/10.2174/1874196702109010001

Sun, Y., Zhuo, C., Wu, G., Wu, Q., Zhang, F., Niu, Z., & Hu, H. (2016). Characteristics of water quality of municipal wastewater treatment plants in China: Implications for resources utilization and management. *Journal of Cleaner Production*, *131*, 1–9. https://doi.org/10.1016/j.jclepro.2016.05.068

Worku, A., Tefera, N., Kloos, H., & Benor, S. (2018). Bioremediation of brewery wastewater using hydroponics planted with vetiver grass in Addis Ababa, Ethiopia. *Bioresources and Bioprocessing*, *5*(39), 1–12. https://doi.org/10.1186/s40643-018-0225-5

Zhu, G., Peng, Y., Li, B., Guo, J., Yang, Q., & Wang, S. (2008). Biological removal of nitrogen from wastewater. *Reviews of Environmental Contamination and Toxicology*, *192*, 159–195. https://doi.org/10.1007/978-0-387-71724-1_5

3 Microalgae Cultivation for Wastewater Treatment and Bioenergy Generation

3.1 INTRODUCTION

The demand for clean energy and water on a daily basis has increased as a consequence of population growth, industrialization, urbanization and other factors affecting the economy and society. As a result, the search for renewable clean energy sources has become increasingly important (Alipoor & Saidi, 2017). Fossil fuels are the primary energy source at the moment; however, they are readily depleted and cannot be easily regenerated over the course of a human lifetime. High usage of fossil fuels is the main driver of greenhouse gas emissions into the atmosphere and climate change. Hence, to fulfil the rising demand for energy and protect the environment, it is essential to prioritize the development of environmentally benign renewable energy sources. The combination of energy and water recovery from municipal or industrial wastewater sources might be a potential solution to these concerns because wastewater is made up of free energy, water and nutrients that can enhance plant growth. Moreover, waste is a resource that may be transformed into useful materials for the manufacturing of new products, source of nutrients or energy. Consequently, the generation and disposal of waste is a worldwide problem that is expected to worsen on all continents (Kaza et al., 2018). The production of waste per year is recorded to amount to 2.01 billion tonnes, while water bodies are estimated to produce about 2 million tonnes of human wastes each day (Unwater, 2014). Effective waste treatment procedures have become increasingly important. Therefore, to achieve ecological sustainability in the area of waste management, a pyramid of waste hierarchy has been developed by the European Waste Framework Directive (EC, 2008). The waste hierarchy is presented in Figure 3.1.

Furthermore, renewable energy sources are predicted to be sufficient to suffice all energy demands of our growing population (Edenhofer et al., 2011). As such, sustainable renewable energy sources including wind, biomass, hydro, geothermal, marine and solar energy are essential for the world's future energy needs (Demirbas, 2008). Biomass is considered to be a unique and attractive green energy feedstock compared to other renewable energy sources due to its abundance in nature, annual regeneration and easy storage (Nansaior et al., 2013). Natural and derived biomass materials are the two types of biomass sources that originate from waste products (Demirbaş,

DOI: 10.1201/9781003359586-3

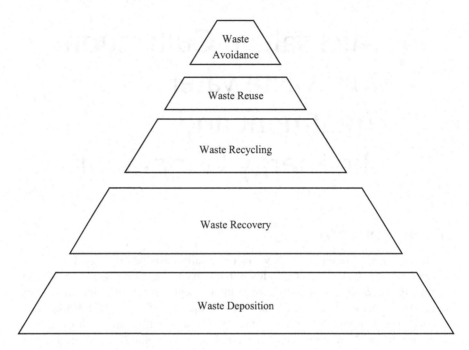

FIGURE 3.1 Pyramid of waste hierarchy.

2001). In other words, biomass is viewed as a solution for climate change mitigation and energy sustainability in developing nations such as Turkey and Malaysia (Bilen et al., 2008).

In addition, biofuel was first introduced by Rudolph diesel in 1900. Simple firewood or more complicated fuels like biodiesel, biojet, bioethanol and bio-oil are examples of biofuels. Currently, biodiesel fuels are utilized in car engines as a straight replacement for diesel fuel or as a component of blends (Demirbaş, 2009). When referring to a blend formulation, the term BXX denotes the percentage of biodiesel that is mixed with fossil fuel. For example, B30 denotes a blend of 70% and 30% diesel and biodiesel, respectively. However, past research has demonstrated that the B20 blend is compatible with most diesel storage and distribution equipment without any need for engine modifications (Balat, 2011), while higher mixes, like B100, are utilized in many engines that are manufactured with little or no modification (Demirbas, 2005). Biodiesel improves combustions and lowers hydrocarbon, soot and carbon monoxide emissions due to the small amount of oxygen present. Biodiesel exhibits better flash point, aromatic content, biodegradability, sulphur content and environmental sustainability properties compared to diesel fuel (Attia & Hassaneen, 2016). However, the price and purity of biodiesel oil are dependent on the kind of feedstock used in the manufacturing procedures. The impact of geography, seasonal crop diversification and the price of crude oil are other considerations. Currently, biodiesel is significantly more expensive than petroleum diesel because of the high cost of the feedstock caused by the necessity for high purity (Mohammed et al., 2014).

3.2 MICROALGAE

Microalgae also called seaweeds are organisms that perform photosynthesis using carbon dioxide, resulting in the production of useful substances known as biomass (Lee & Lavoie, 2013). It has been estimated that there are about 150,000 identified microalgae species grouped into eight phyla: Xanthophyta, Chlorophyta, Phaeophyta, Diatom, Euglenophyta, Pyrrophyta, Rhodophyta and Chrysophyta. They usually develop in nutrient-rich marine environment such as fresh and marine water, municipal wastewater (Zhou et al., 2012), industrial wastewater and animal wastewater (Hu et al., 2013) that contains enough quantity of carbon, nitrogen and phosphorous.

The future market of algae-derived products is currently trending upwards. Additionally, Mehta et al. (2018) postulated that by the year 2024, the global value for microalgae products would be worth about 1143 million US dollars. Furthermore, *Chlorella vulgaris* and *Botryococcus braunii* algal species contain high amounts of lipids which indicates a strong potential for biodiesel production (Najafi et al., 2011). According to Chisti (2007), microalgae are the sole modern renewable source of biofuels and therefore are anticipated to be significantly less harmful to the environment and the global food supply chain than traditional biofuel-producing crops. Microalgae are more suitable for biofuel than lignocellulosic materials due to their inherent high lipid content, semi-steady state production and suitability in a variety of climates (Miao et al., 2004). Microalgal biofuels are distinguished by high caloric value, low viscosity and low density, making them a better biofuel when compared to plant biofuels (Clarens et al., 2010).

Microalgae can be cultivated in closed, open or hybrid systems for the production of feedstock or biomass. Open systems are the most cost-effective, but they have the drawback of having little ability to regulate contamination from predators, whereas photobioreactors (PBR) may be employed in closed systems. In other words, PBR offers a beneficial method for controlling culture factors including carbon dioxide, temperature, light and pH with high financial investment. In contrast, the hybrid systems integrate the closed and open systems in a two-stage cultivation process that accommodates the culturing of the inoculum using photobioreactors in the initial stage, followed by cultivation of the algae in the second stage using open ponds. The type of technique to be employed is determined by the planned use of the biomass and the environmental parameters of the production site (Aratboni et al., 2019). Despite the benefits of microalgae-derived biofuels over other higher plants, the production of biofuels from microalgae cannot account for a sizeable portion of the global supply of liquid fuels due to the numerous obstacles that hinder the processes. In this respect, different techniques and instruments such as high-rate algal ponds, photobioreactors and microfuel cells have been used for effective microalgae cultivation for treatment of wastewater and biomass recovery, which is further discussed in the next section.

In this regard, several approaches and equipment including high-rate algal ponds, photobioreactors and microfuel cells have been employed for successful microalgae culture for wastewater treatment and biomass recovery. These techniques are covered in more detail in the following sections.

TABLE 3.1

Overview of Some Work That Has Been Carried Out in This Area

Author	Research Work	Main Discoveries
Baudelet et al. (2017)	Cultivation of different Chlorophyta (green algae) species.	*Chlorella protothecoides, Chlorella ellipsoidea* and *Chlorella kessleri* are good candidates for small-scale phytoremediation and biofuel generation.
Ren et al. (2017)	Cultivation of *C. vulgaris* for wastewater treatment and biomass generation using lab-scale photobioreactor.	The photobioreactor was conducive for the microalgae growth and nutrient removal from the wastewater.
Ansari et al., 2017	Microalgae cultivation for carbon dioxide sequestration and biofuel production.	The local microalgae strain was observed to polish the wastewater within 7 days retention time.
Martín-Juárez et al. (2019)	Optimization of fermentable monosaccharides from algal biomass grown in photobioreactors for wastewater treatment.	The process improved the solubilization up to 97%, while the degradation remained constant.
de Souza Leite et al. (2019)	Cultivated microalgae in municipal and piggery wastewater collected from Brazil.	The result revealed that NH_3 was removed through air stripping, which decreased the N:P ratio during cultivation. This has seriously affected the biomass productivity and nutrient uptake.

3.3 HIGH-RATE ALGAL PONDS (HRAPs)

High-rate algal ponds (HRAPs) can be described as shallow raceway ponds used for microalgae cultivation and wastewater treatment (Craggs et al., 2014). HRAPs are gaining rekindled attention due to their ability to treat wastewater with low cost and low energy than typical activated sludge systems and photobioreactors (Garfí et al., 2017). These systems enhance microalgal biomass generation that can be used as feedstock biofuel and valuable bioproducts production (Young et al., 2017). It is worth noting that they can be easily deployed in areas where environmental conditions are favourable for optimal microalgae development. Additionally, HRAP is a natural system that promotes sustainable technique for wastewater treatment due to its potential in lowering greenhouse gas emissions during nutrient uptake (Acién, 2016). However, enormous surface area required for effective wastewater treatment and biomass productivity is a key barrier to the adoption and proliferation of HRAP technology (Acién, 2016). Hence, it is necessary to redesign the maintenance and operational systems of HRAPs by eliminating the primary treatment from the entire process. Furthermore, the potentials of HRAP technology in remediation of different sources of wastewater and bioenergy generation have been the focus of recent research.

Arashiro et al. (2019) evaluated the efficacy of wastewater treatment and biomass generation using HRAPs. High-rate algal ponds with treatment (HRAP-PT) and high-rate algal ponds without treatment (HRAP-noPT) configurations was used for the study. Water quality of the wastewater samples including turbidity, temperature, dissolved oxygen (DO), pH, total suspended solids (TSS), chlorophyll-a and volatile suspended solids (VSS) was assessed. The results obtained demonstrated that there were no appreciable changes in nutrient uptake of the HRAP-PT and HRAP-noPT in terms of wastewater treatment. In addition, the difference between the two systems depended more on the inert materials introduced into the HRAP-noPT. Additionally, the outcome reported by Arashiro et al. (2019) supports the theory put out by Posadas et al. (2017), which claimed that, based on a theoretical analysis, primary treatment utilizing the HRAP technology is not necessary in municipal wastewater treatment. Therefore, the elimination of primary treatment prior to an HRAP will not influence the efficiency of wastewater treatment, rather it would minimize maintenance, lower expenses and footprint in HRAP systems.

3.4 PHOTOBIOREACTORS (PBRs) AND MEMBRANE BIOREACTORS (MBRs)

The type of reactor utilized during the growing process is one of the main variables that affects the productivity of microalgae biomass. Different kinds of photobioreactors are available for growing microalgae. Photobioreactors are classified as tabular, column, membrane and flat plate photobioreactors (Moreno-Garcia et al., 2017). In comparison to open systems, microalgae cultivation in photobioreactors involves higher capital expenditures for lighting, circulation system, culture medium and CO_2 feed, although it is easier to maintain due to its low contamination rate and better productivity compared to the open systems.

Recently, there has been a lot of attention in the use of membrane bioreactors (MBRs) for the treatment of wastewater (Di Bella et al., 2018; Mannina et al., 2017a). MBR applies biological processes that incorporate perm-selective membrane for final solid-liquid separation. Generally, MBR offers excellent effluent quality that is characterized with volumetric loading rate, low footprint and high sludge retention time values, and they are excellent in solid-liquid separations in wastewater treatment (Collivignarelli et al., 2016). Additionally, MBRs have good capacity for solid retention, which allows operation with higher biomass concentration compared with conventional sludge activation system, resulting in the ability to use lower volume reactors and minimal footprint requirement. Nonetheless, membrane fouling is the major drawback associated with MBR operations (Cosenza et al., 2013; Xiong et al., 2016). High energy consumption caused by membrane aeration is also a disadvantage of MBRs (Mannina et al., 2017b). Recently, Di Trapani et al. (2019) used membrane bioreactors for the treatment of wastewater. The findings demonstrated that total resistance to filtration was observed, which is probably related to fast deposition of particles on the membrane surface. This outcome might be attributed to granule instability, which occurs when the average size of crushed granules decreases, shortening the membrane life. Moreover, the quality of the membrane permeate would fulfil water reuse criteria, supporting a cleaner and more sustainable wastewater treatment technology.

3.5   MICROBIAL FUEL CELL (MFC)

Microbial fuel cell is a device that harnesses respiring bacteria to directly transform organic substrates into electrical energy. This device is one of the emerging technologies that simultaneously removes organic matters in wastewater and generates power. MFCs have been used to treat low-strength wastewater with low footprint (Rittmann, 2008; Watanabe, 2008). MFC is made up of a cathode, electrolyte medium and anode, microorganisms and proton exchange membrane (PEM). The microorganisms are found in the anode compartment of the MFC, and they utilize organic substrates as fuel to generate protons and electrons. Nicotinamide adenine dinucleotide (NADH) transports these electrons (Choudhury et al., 2017). Furthermore, MFCs offer several advantages for treatment of wastewater such that no energy for aeration is required, power is generated from pollutants, there is a possible reduction in waste sludge production and the growth of active microorganisms is promoted while treating wastewater. Hence, they have appreciable functional stability under ambient temperature (Moon et al., 2005). MFCs have various drawbacks, such as reliance on biofilms for electron transfer, whereas anaerobic digesters do not have this problem since they effectively reuse the microbial community without cell immobilization (Pham et al., 2006). Additionally, large contact area is required for biofilm to accumulate on the anode when MFCs are employed for wastewater treatment. Furthermore, low catalytic rate of the microorganisms is another disadvantage of using MFC (Du et al., 2007). Meanwhile, the anaerobic environment in the MFC anodic chamber does not allow for efficient nutrient removal. As a result, an MFC that can quickly aid in nutrient uptake during wastewater treatment is required. This will make MFC promising for future deployment (Choudhury et al., 2017). Previous research has shown that MFC is effective for removing organic materials from wastewater within a suitable retention time. Ishii et al. (2008) showed that a mixed culture from sewage sludge could generate 22% more energy than *Geobacter sulfurreducens* in culture in a single-chamber MFC with an air cathode. Further, pure culture is more effective than mixed culture in understanding the mechanics of electricity production and developing strategies to improve process parameters, although mixed culture is better than pure culture in wastewater treatment because they utilize different type of substrates (Chaudhuri & Lovley, 2003). Furthermore, the bacteria utilized in MFC should be electrochemically active and capable of withstanding the reaction temperature (Rabaey & Verstraete, 2005). Other bacteria including *Desulfobulbus propionicus*, *Klebsiella pneumonia* and *Rhodopseudomonas palustris* demonstrate prospect for power generation in MFCs (Sharma & Kundu, 2010). Future study in this area should focus on screening, microbial identification and their capacity for wastewater treatment and bioenergy generation.

3.6 CONCLUSION

Researchers have extensively examined and reported on the culturing of microalgae for wastewater treatment procedures and bioenergy generation. It can be deduced that these biological techniques are feasible for large-scale integration of wastewater remediation and energy production, if the aforementioned limitations surrounding HRAPs, photobioreactors and MFCs are overcome. These limitations have led to the application of microalgae for simultaneous wastewater remediation and bioenergy production tedious and less attractive to a certain level that has resulted in making the procedures expensive, uncompetitive with the conventional fossil fuels and a drawback for using microalgae as sustainable renewable aquatic plant. Invariably, there is the need to find other alternatives through exploring the potentials of other aquatic plant species such as giant salvinia, water hyacinth and water lettuce for wastewater treatment and bioenergy generation to meet up with the rising demand for clean water and development of sustainable renewable energy sources.

REFERENCES

Acién, F. G. (2016). Wastewater treatment using microalgae: How realistic a contribution might it be to significant urban wastewater treatment? *Applied Microbiology and Biotechnology.* https://doi.org/10.1007/s00253-016-7835-7

Alipoor, A., & Saidi, M. H. (2017). Numerical study of hydrogen-air combustion characteristics in a novel micro-thermophotovoltaic power generator. *Applied Energy, 199*, 382–399. https://doi.org/10.1016/j.apenergy.2017.05.027

Ansari, A. A., Khoja, A. H., Nawar, A., Qayyum, M., & Ali, E. (2017). Wastewater treatment by local microalgae strains for CO2 sequestration and biofuel production. *Applied Water Science, 7*(7), 4151–4158. https://doi.org/10.1007/s13201-017-0574-9

Arashiro, L. T., Ferrer, I., Rousseau, D. P. L., Van Hulle, S. W. H., & Garfí, M. (2019). The effect of primary treatment of wastewater in high rate algal pond systems: Biomass and bioenergy recovery. *Bioresource Technology, 280*(January), 27–36. https://doi.org/10.1016/j.biortech.2019.01.096

Aratboni, H. A., Rafiei, N., Granados, R. G., Alemzadeh, A., & Morones-Ramírez, R. J. (2019). Biomass and lipid induction strategies in microalgae for biofuel production and other applications. *Microbial Cell Factories, 18*(178), 1–17. https://doi.org/10.1186/s12934-019-1228-4

Attia, A. M. A., & Hassaneen, A. E. (2016). Influence of diesel fuel blended with biodiesel produced from waste cooking oil on diesel engine performance. *Fuel, 167*(November), 316–328. https://doi.org/10.1016/j.fuel.2015.11.064

Balat, M. (2011). Potential alternatives to edible oils for biodiesel production - A review of current work. *Energy Conversion and Management, 52*(2), 1479–1492. https://doi.org/10.1016/j.enconman.2010.10.011

Baudelet, P. H., Ricochon, G., Linder, M., & Muniglia, L. (2017). A new insight into cell walls of Chlorophyta. *Algal Research, 25*(July), 333–371. https://doi.org/10.1016/j.algal.2017.04.008

Bilen, K., Ozyurt, O., Bakirci, K., Karsli, S., Erdogan, S., Yilmaz, M., & Comakli, O. (2008). Energy production, consumption, and environmental pollution for sustainable development: A case study in Turkey. *Renewable and Sustainable Energy Reviews, 12*(6), 1529–1561. https://doi.org/10.1016/j.rser.2007.03.003

Chaudhuri, S. K., & Lovley, D. R. (2003). Electricity generation by direct oxidation of glucose in mediatorless microbial fuel cells. *Nature Biotechnology, 21*(10), 1229–1232. https://doi.org/10.1038/nbt867

Chisti, Y. (2007). Biodiesel from microalgae. *Biotechnology Advances, 25*(3), 294–306. https://doi.org/10.1016/j.biotechadv.2007.02.001

Choudhury, P., Shankar, U., Uday, P., Mahata, N., & Nath, O. (2017). Performance improvement of microbial fuel cells for waste water treatment along with value addition: A review on past achievements and recent perspectives. *Renewable and Sustainable Energy Reviews, 79*(May), 372–389. https://doi.org/10.1016/j.rser.2017.05.098

Clarens, A. F., Resurreccion, E. P., White, M. A., & Colosi, L. M. (2010). Environmental life cycle comparison of algae to other bioenergy feedstocks. *Environmental Science & Technology, 44*(5), 1813–1819.

Collivignarelli, M. C., Abb, A., Castagnola, F., & Bertanza, G. (2016). Minimization of municipal sewage sludge by means of a thermophilic membrane bioreactor with intermittent aeration. *Journal of Cleaner Production*, 1–8. https://doi.org/10.1016/j.jclepro.2016.12.101

Cosenza, A., Di Bella, G., Mannina, G., & Torregrossa, M. (2013). The role of EPS in fouling and foaming phenomena for a membrane bioreactor. *Bioresource Technology*. https://doi.org/10.1016/j.biortech.2013.08.026

Craggs, R., Park, J., Heubeck, S., & Sutherland, D. (2014). High rate algal pond systems for low-energy wastewater treatment, nutrient recovery and energy production. *New Zealand Journal of Botany, 52*(1), 60–73. https://doi.org/10.1080/0028825X.2013.861855

de Souza Leite, L., Hoffmann, M. T., & Daniel, L. A. (2019). Microalgae cultivation for municipal and piggery wastewater treatment in Brazil. *Journal of Water Process Engineering, 31*(December 2018), 1–7. https://doi.org/10.1016/j.jwpe.2019.100821

Demirbaş, A. (2001). Biomass resource facilities and biomass conversion processing for fuels and chemicals. *Energy Conversion and Management, 42*(11), 1357–1378. https://doi.org/10.1016/S0196-8904(00)00137-0

Demirbas, A. (2005). Biodiesel production from vegetable oils via catalytic and non-catalytic supercritical methanol transesterification methods. *Progress in Energy and Combustion Science, 31*(5–6), 466–487. https://doi.org/10.1016/j.pecs.2005.09.001

Demirbas, A. (2008). Biofuels sources, biofuel policy, biofuel economy and global biofuel projections. *Energy Conversion and Management, 49*(8), 2106–2116. https://doi.org/10.1016/j.enconman.2008.02.020

Demirbaş, A. (2009). Production of biodiesel from algae oils. *Energy Sources, Part A: Recovery, Utilization and Environmental Effects, 31*(2), 163–168. https://doi.org/10.1080/15567030701521775

Di Bella, G., Di Trapani, D., & Judd, S. (2018). Separation and purification technology fouling mechanism elucidation in membrane bioreactors by bespoke physical cleaning. *Separation and Purification Technology, 199*(October 2017), 124–133. https://doi.org/10.1016/j.seppur.2018.01.049

Di Trapani, D., Corsino, S. F., Torregrossa, M., & Viviani, G. (2019). Treatment of high strength industrial wastewater with membrane bioreactors for water reuse: Effect of pretreatment with aerobic granular sludge on system performance and fouling tendency. *Journal of Water Process Engineering, 31*(May), 100859. https://doi.org/10.1016/j.jwpe.2019.100859

Du, Z., Li, H., & Gu, T. (2007). A state of the art review on microbial fuel cells: A promising technology for wastewater treatment and bioenergy. *Biotechnology Advances*, *25*, 464–482. https://doi.org/10.1016/j.biotechadv.2007.05.004

Edenhofer, O., Pichs-Madruga, R., Sokona, Y., Seyboth, K., Eickemeier, P., Matschoss, P., Hansen, G., Kadner, S., Schlömer, S., Zwickel, T., & Von Stechow, C. (2011). IPCC, 2011: Summary for Policymakers. In *IPCC Special Report on Renewable Energy Sources and Climate Change Mitigation*. Cambridge University Press, United Kingdom. https://doi.org/10.5860/CHOICE.49-6309

European Commission (EC). (2008). Directive 2008/98/EC of the European Parliament and of the Council of 19 November 2008 on waste. *Official Journal of the European Union*, *312*(13), 3–30.

Garfí, M., Flores, L., & Ferrer, I. (2017). Life cycle assessment of wastewater treatment systems for small communities: Activated sludge, constructed wetlands and high rate algal ponds. *Journal of Cleaner Production*, *161*, 211–219. https://doi.org/10.1016/j.jclepro.2017.05.116

Hu, B., Zhou, W., Min, M., Du, Z., Chen, P., Ma, X., Liu, Y., Lei, H., Shi, J., & Ruan, R. (2013). Development of an effective acidogenically digested swine manure-based algal system for improved wastewater treatment and biofuel and feed production. *Applied Energy*, *107*(February 2020), 255–263. https://doi.org/10.1016/j.apenergy.2013.02.033

Ishii, S., Watanabe, K., Yabuki, S., Logan, B. E., & Sekiguchi, Y. (2008). Comparison of electrode reduction activities of Geobacter sulfurreducens and an enriched consortium in an air-cathode microbial fuel cell. *Applied and Environmental Microbiology*, *74*(23), 7348–7355. https://doi.org/10.1128/AEM.01639-08

Kaza, S., Yao, L. C., Bhada-Tata, P., & van Woerden, F. (2018). What a Waste 2.0: A global snapshot of solid waste management to 2050. *Syria Studies*, *7*(1). https://www.researchgate.net/publication/269107473_What_is_governance/link/548173090cf22525dcb61443/download%0Ahttp://www.econ.upf.edu/~reynal/Civil wars_12December2010.pdf%0A https://think-asia.org/handle/11540/8282%0Ahttps://www.jstor.org/stable/41857625

Lee, R. A., & Lavoie, J. M. (2013). From first- to third-generation biofuels: Challenges of producing a commodity from a biomass of increasing complexity. *Animal Frontiers*, *3*(2), 6–11. https://doi.org/10.2527/af.2013-0010

Mannina, G., Capodici, M., Cosenza, A., & Di Trapani, D. (2017a). Greenhouse gas emissions and the links to plant performance in a fixed-film activated sludge membrane bioreactor – pilot plant experimental evidence. *Bioresource Technology*. https://doi.org/10.1016/j.biortech.2017.05.043

Mannina, G., Capodici, M., Cosenza, A., Di Trapani, D., & George, A. (2017b). The effect of the solids and hydraulic retention time on moving bed membrane bioreactor performance. *Journal of Cleaner Production*. https://doi.org/10.1016/j.jclepro.2017.09.200

Martín-Juárez, J., Vega-Alegre, M., Riol-Pastor, E., Muñoz-Torre, R., & Bolado-Rodríguez, S. (2019). Optimisation of the production of fermentable monosaccharides from algal biomass grown in photobioreactors treating wastewater. *Bioresource Technology*, *281*(December 2018), 239–249. https://doi.org/10.1016/j.biortech.2019.02.082

Mehta, P., Singh, D., Saxena, R., & Rani, R. (2018). High-Value Coproducts from Algae— An Innovative Way to Deal with Advance Algal Industry. In R. R. Singhania, R. A. Agarwal, R. Praveen Kumar, R. K. Sukumaran (Eds.), *Waste to Wealth*, Springer Nature, Singapore. https://doi.org/10.1007/978-981-10-7431-8

Miao, X., Wu, Q., & Yang, C. (2004). Fast pyrolysis of microalgae to produce renewable fuels. *Journal of Analytical and Applied Pyrolysis, 71*, 855–863. https://doi.org/10.1016/j. jaap.2003.11.004

Mohammed, N. I., Kabbashi, N. A., Alam, M. Z., & Mirghani, M. E. (2014). Jatropha Curcas Oil Characterization and Its Significance for Feedstock Selection in Biodiesel Production. *5th International Conference on Food Engineering and Biotechnology, 65*, 6. https://doi.org/10.7763/IPCBEE

Moon, H., Chang, I. S., Jang, J. K., & Kim, B. H. (2005). Residence time distribution in microbial fuel cell and its influence on COD removal with electricity generation. *Biochemical Engineering Journal, 27*, 59–65. https://doi.org/10.1016/j.bej.2005.02.010

Moreno-Garcia, L., Adjallé, K., Barnabé, S., & Raghavan, G. S. V. (2017). Microalgae biomass production for a biorefinery system: Recent advances and the way towards sustainability. *Renewable and Sustainable Energy Reviews, 76*(January), 493–506. https://doi.org/10.1016/j.rser.2017.03.024

Najafi, G., Ghobadian, B., & Yusaf, T. F. (2011). Algae as a sustainable energy source for biofuel production in Iran: A case study. *Renewable and Sustainable Energy Reviews, 15*(8), 3870–3876. https://doi.org/10.1016/j.rser.2011.07.010

Nansaior, A., Patanothai, A., Rambo, A. T., & Simaraks, S. (2013). The sustainability of biomass energy acquisition by households in urbanizing communities in Northeast Thailand. *Biomass and Bioenergy, 52*, 113–121. https://doi.org/10.1016/j.biombioe.2013.03.011

Pham, B. T. H., Rabaey, K., Aelterman, P., Clauwaert, P., De Schamphelaire, L., Boon, N., & Verstraete, W. (2006). Microbial fuel cells in relation to conventional anaerobic digestion technology. *Engineering in Life Science, 3*, 285–292. https://doi.org/10.1002/ elsc.200620121

Posadas, E., Muñoz, R., & Guieysse, B. (2017). Integrating nutrient removal and solid management restricts the feasibility of algal biofuel generation via wastewater treatment. *Algal, 22*, 39–46. https://doi.org/10.1016/j.algal.2016.11.019

Rabaey, K., & Verstraete, W. (2005). Microbial Fuel Cells: Novel Biotechnology for Energy Generation, *23*(6). https://doi.org/10.1016/j.tibtech.2005.04.008

Ren, H., Tuo, J., Addy, M. M., Zhang, R., Lu, Q., Anderson, E., Chen, P., & Ruan, R. (2017). Cultivation of Chlorella vulgaris in a pilot-scale photobioreactor using real centrate wastewater with waste glycerol for improving microalgae biomass production and wastewater nutrients removal. *Bioresource Technology, 245*(July), 1130–1138. https://doi.org/10.1016/j.biortech.2017.09.040

Rittmann, B. E. (2008). Opportunities for renewable bioenergy using microorganisms. *Biotechnology and Bioengineering, 100*(2), 203–212. https://doi.org/10.1002/bit.21875

Sharma, V., & Kundu, P. P. (2010). Enzyme and microbial technology biocatalysts in microbial fuel cells. *Enzyme and Microbial Technology, 47*(5), 179–188. https://doi.org/10.1016/j. enzmictec.2010.07.001

Unwater. (2014). United Nations inter-agency coordination mechanism for all freshwater and Sanitation. http://www.unwater.org/statistics/statistics-detail/en/c/211801/2014

Watanabe, K. (2008). Recent developments in microbial fuel cell technologies for sustainable bioenergy. *Journal of Bioscience and Bioengineering, 106*(6), 528–536. https://doi. org/10.1263/jbb.106.528

Xiong, J., Fu, D., Singh, R. P., & Ducoste, J. J. (2016). Structural characteristics and development of the cake layer in a dynamic membrane bioreactor. *Separation and Purification Technology.* https://doi.org/10.1016/j.seppur.2016.04.040

Young, P., Taylor, M., & Fallowfield, H. J. (2017). Mini-review: High rate algal ponds, flexible systems for sustainable wastewater treatment. *World Journal of Microbiology and Biotechnology, 33*(17), 13. https://doi.org/10.1007/s11274-017-2282-x

Zhou, W., Min, M., Li, Y., Hu, B., Ma, X., Cheng, Y., Liu, Y., Chen, P., & Ruan, R. (2012). A hetero-photoautotrophic two-stage cultivation process to improve wastewater nutrient removal and enhance algal lipid accumulation. *Bioresource Technology, 110*, 448–455. https://doi.org/10.1016/j.biortech.2012.01.063

4 Cultivation of Aquatic Plants for Wastewater Treatment

4.1 INTRODUCTION

Pollution is among the core health-threatening issues affecting our environment. Natural water sources have been negatively affected by organic and inorganic pollutants from agricultural, industrial and domestic waste (Dipu et al., 2010). Environmental problems in the atmosphere are a result of organic and inorganic soluble found in wastewater which leads to water pollution (Thongtha et al., 2014). Similarly, water pollution has a detrimental effect on agriculture, resulting in bioaccumulation of toxic metals into the food chain (George & Gabriel, 2017). Hence, effective wastewater treatment before discharge into the environment is critical for preventing water pollution (Daverey et al., 2019). Furthermore, macrophytes (aquatic plants) have the capacity to absorb pollutants including heavy metals, organic and inorganic, nanoparticles, radioactive elements, nanoparticles and pharmaceutical compounds in industrial, agricultural and domestic wastewater sources either at secondary or tertiary phytoremediation processes. This could be through rhizofiltration, phytoextraction, phytostabilization, phytovolatilization, phytodegradation or phytotransformation. As a result, we need to improve the quality of treated water to meet up with international standard requirements. Hence, the combination of energy and recovery of water from domestic or industrial wastewater sources through cultivation of aquatic plants could be a possible solution to tackle problems of water scarcity and the depleting fossil fuels.

4.2 PHYTOREMEDIATION OF WASTEWATER

Phytoremediation techniques have been used for decades. In the late 19th century, yellow calamine violet, *Thlaspi caerulescens*, *Viola calaminaria* and *Alpine pennycress* were species of wild flower plants reported to accumulate large quantities of metals (Sharma & Pandey, 2014). Phytoremediation is an innovative technique that utilizes plant roots in degradation of pollutants in soils, sludge, sediments and water (Wickramasinghe & Jayawardana, 2018). Additionally, plant survival is influenced by the condition and toxicity of the contaminated soil. Bioremediation systems based on plants have gained interest as potential strategies for cleaning up contaminated soil and water. Numerous plant species have demonstrated considerable prospects as phytoremediation agents. Among these plants are grasses, trees and a variety of monocots and dicots (Raju et al., 2010).

Besides, the construction of hydroponic systems for the cultivation of aquatic plants meets all prerequisites of environmentally friendly methods for wastewater

DOI: 10.1201/9781003359586-4

treatment (Almuktar et al., 2018). Additionally, aquatic plant cultivation for wastewater purification processes is more appropriate in countries with warm temperatures. Due to these benefits, it has the potential to be an ideal option for developing countries (Magwaza et al., 2020). In addition, aquatic plants are more effective in wastewater treatment than terrestrial plants, owing to their higher pollutant absorbing capabilities and fast growth (Singh et al., 2012). Furthermore, the potentials of aquatic plants in treating wastewater have been investigated and reviewed extensively by many authors (Dhir et al., 2009; Mahmood et al., 2005; Wickramasinghe & Jayawardana, 2018). Moreover, phytoremediation methods for wastewater treatment can be operated under three principles: (1) selection and cultivation of good aquatic plant candidate; (2) absorption and degradation of the excess nutrients by the cultivated plants; and (3) harvesting and useful application of plant biomass generated by the remediation system (Lu et al., 2010). Else the decomposing plant tissue will release the stored nutrients back into the environment (Ting et al., 2018). However, phytoremediation is limited to the depth and surface area occupied by the roots, slow growth and the increasing time required for biomass generation. Additionally, it is not possible to completely prevent contaminant leaching into groundwater using plant-based remediation systems. The toxicity of contaminated land and the general condition of the soil have an impact on plant survival.

4.3   MECHANISMS OF PHYTOREMEDIATION

The mechanisms of phytoremediation can be categorized into phytodegradation, phytoextraction, phytostabilization, phytovolatilization and rhizofiltration (Ting et al., 2018). The description of each type of mechanism is presented in Table 4.1 and Figure 4.1. Additionally, Table 4.2 outlines the advantages and drawbacks of phytoremediation methods used in the treatment of wastewater.

4.4   THE ROLES OF AQUATIC PLANTS IN PHYTOREMEDIATION OF WASTEWATER

The effectiveness of macrophytes in eradicating contaminants is dependent on the concentration of the pollutants, retention time, plant features (root system and species) and environmental factors (temperature, pH) (Anand et al., 2017). Besides, submerged plants (*Ruppia maritima, Egeria densa, Myriophyllum aquaticum, Najas marina, Hygrophila corymbosa, Vallisneria americana, Hydrilla verticillata* and *Najas marina*), emergent plants (*Typha* spp., *Justicia americana, Distichlis spicata, Phragmites australis, Imperata cylindrical, Cyperus* spp., *Nymphaea* spp., *Hydrochloa caroliniensis, Nuphar lutea, Iris virginica* and *Diodia virginiana*) and free-floating macrophytes (*Riccia fluitans, Spirodela polyrhiza, Pistia stratiotes, Landoltia punctata, Salvinia molesta, Marsilea mutica, Lemna* spp., *Azolla pinnata* and *Eichhornia crassipes*) have been utilized in remediation of wastewater (Ekperusi et al., 2019; USDA, 2019). In this context, the available records on macrophyte species that have been used in phytoremediation of agricultural, industrial and domestic wastewater are presented in Tables 4.3 and 4.4.

TABLE 4.1

Mechanisms of Phytoremediation

Techniques	Mechanisms	Advantages
Phytodegradation	Plants and microorganisms either convert or destroy organic pollutants into non-toxic compounds within the rhizosphere or plant tissues.	Waste reduction and biomass generation for bioenergy production.
Phytoextraction	It is a natural process through which plants absorb, translocate and accumulate organic and inorganic substances from the environment. Plant roots absorb pollutants along with nutrients and water. The pollutants are not destroyed but stored in the leaves and shoots of the plants.	The removal of contaminants is economical and permanent.
Phytostabilization	Heavy metals and other contaminants are immobilized in the soil via sorption by roots, precipitation and complexation instead of decomposition.	Prevents soil erosion, and no need for disposal of hazardous waste.
Phytovolatilization	Contaminants are absorbed by plants and transpired via the xylem, where they are further converted into volatile forms and released into the environment as gaseous chemicals through the leaves. Phytovolatilization method can be used to remove heavy metals and organic contaminants from the environment.	Pollutants can be directly discharged into the atmosphere.
Rhizofiltration	Pollutants are absorbed through the root surface, precipitated in the root zone or other plant components.	Pollutants are retained in the roots after in situ and ex situ application.

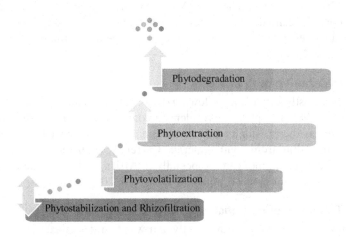

FIGURE 4.1 Mechanisms of phytoremediation.

TABLE 4.2
Advantages and Drawbacks of Phytoremediation in Wastewater

Advantages	Drawback
Requires low capital and energy (Ahmad et al., 2017)	Phytoremediation is limited to shallow ponds (Ahmad et al., 2017)
Environmental friendliness (Parnian et al., 2016)	Toxicity of contaminants to plants (phytotoxicity) (Ahmad et al., 2017)
Utilizes natural and renewable source (Ng et al., 2017)	Requires longer time compared to conventional methods (Ahmad et al., 2017)
Less secondary waste generation (Parnian et al., 2016)	Plant mats serve as a breeding ground for harmful insects such as mosquitoes
Less carbon footprint (Anand et al., 2017)	Tendency of the pollutants to infiltrate the food chain (Ahmad et al., 2017)
Wastewater reclamation and nutrient recovery (Anand et al., 2017)	
Biomass generation for use as raw material (Parnian et al., 2016)	

4.5 APPLICATIONS OF *P. STRATIOTES, S. MOLESTA* AND *E. CRASSIPES* IN PHYTOREMEDIATION OF WASTEWATER

4.5.1 *PISTIA STRATIOTES* PLANTS

Pistia stratiotes also referred to as water lettuce is among the most abundant free-floating macrophytes (Neuenschwander et al., 2009). The genus name was originated from the Latin word "pistos", which translates as "water" and alludes to the aquatic characteristics of the plant. It proliferates in contaminated water with high nutrient content. *P. stratiotes* grows at a similar rate as *E. crassipes* (60–110 t/ha/year) and have a proclivity for weedy behaviour (Mishima et al., 2008). Additionally, *P. stratiotes* is a potential candidate for wastewater phytoremediation due to their availability, low-cost harvesting and capability to tolerate temperatures ranging from 21°C to 30°C (Aswathy, 2017).

4.5.1.1 Distribution of *P. stratiotes*

P. stratiotes is mostly found in slow-flowing rivers, ponds, streams or lakes (Akobundu & Agyakwa, 1998). The plant is prevalent in continents, excluding Antarctica and Europe. There is the possibility that the original spread occurred through ballast water on ships sailing from South America. Currently, *P. stratiotes* is the most distributed macrophytes in the world, especially in Africa, and is commercially available as an ornament for ponds (Labrada & Fornasari, 2002).

4.5.1.2 Taxonomy of *P. stratiotes*

P. stratiotes is among the Araceae family and was first described by three ancient Egyptians and two Greek philosophers. Presently, monospecific *P. stratiotes* is the sole representative of the *Pistiodea* subfamily in Araceae. The diversity of this tax-onomically unique species, which is the only free-floating aroid, is shown by the many synonyms and outdated sub-specific names (Neuenschwander et al., 2009). The detailed taxonomy of *P. stratiotes* is presented in Table 4.5.

4.5.1.3 Description of *P. stratiotes*

P. stratiotes is a freshwater herb characterized by dense, succulent buds, thick sto-lon, translucent, unbranched and fibrous roots. The plants are round in shape with height and width of 6–12 and 5 cm, respectively. They float on the surface of water, with their roots suspended underneath the floating leaves. Additionally, they are pale green in colour, with parallel veins and wavy edges, and are coated in small hairs that create basket-like structures that trap air bubbles, hence increasing the resis-tance of the plants. The flowers are covered in the centre of the plants and are dioe-cious. Upon successful fertilization, small green berries developed (Akobundu & Agyakwa, 1998).

4.5.1.4 Growth of *P. stratiotes*

P. stratiotes is an evergreen perennial plant that naturally forms large floating colo-nies in tropical and subtropical waterways (Lima et al., 2013). The temperature resis-tance of *P. stratiotes* ranges from 15°C to 35°C, while the maximum temperature ranges from 22°C to 30°C and pH of 6.5–7.2 (Rivers, 2002). However, *P. stratiotes* is vulnerable to frost and cannot resist salt concentrations of 1.66% and above (CABI, 2019). *P. stratiotes* exhibits asexual reproduction, which is prone to breakage and propagation of the buds, seeds or stolon. The seeds are easily transported by water across great distances as they float during the first 2 days of maturity. A short stolon connects the mother and daughter plants, forming thick mats. Typically, new devel-opment (growth) usually occurs with the formation of smaller daughter plants con-nected to the original growth by stolon (Labrada & Fornasari, 2002). The diagram displaying the leaves of *P. Stratiotes* plant is presented in Figure 4.2.

4.5.1.5 Efficiency of *P. stratiotes* in Phytoremediation of Wastewater

Several research works have been published on applications of macrophytes in the treatment of wastewater (Kochi et al., 2020). Only recent research (2015 and beyond) that employed macrophytes in wastewater treatment were selected for comparison in this review. As a result, it was found that some authors identified specific plants in their study. In contrast, other researchers worked with a variety of plants. Thus, this subsection covers past research and contributions on the efficiency of *P. stratiotes*, *S. molesta* and *E. crassipes* in wastewater treatment. The research performed by Mukherjee et al. (2015) was restricted to phytoremediation of parboiled rice mill wastewater using batch study, and only four physicochemical tests (COD, NH_4-N, NO_3-N and P) were performed for 15 days. The retention time was not defined in the study. Hanks et al. (2015) focused only on the removal of silver nanoparticles (AgNP) from contaminated water. The design and dimension of the treatment sys-tems was not described. Meanwhile, Nivetha et al. (2016) employed plastic trough with different dimensions for cultivation of the plants for 10 days at 2 days detention time. Only three physicochemical tests (NH_4-N, PO_4^{3-} and NO_3^{-}) was performed in

TABLE 4.3
Macrophytes Used in Phytoremediation of Domestic and Agricultural Wastewater

Macrophytes	Domestic/ Agricultural Wastewater	Pollutant/Nutrient Reduction	Duration of Sample Study	Researchers
Water hyacinth	Municipal wastewater	Nitrate (92%), NH$_3$ (81%), Phosphorous (67%), COD (49%).	24 days/ reactor tanks.	Kutty et al. (2009)
Hydrilla verticillata	Domestic wastewater	NO$_3$-N (50.4%), orthophosphate (52.6%), electrical conductivity (15.7%), NH$_3$-N (39.5%), COD (36.1%), TSS (15.8%), DO (140.1%), phosphate (44.7%), Mg (37.2%), Ca (24.4%), Ca hardness (25%).	1 year/ artificial culture system	Patel and Kanungo (2012)
Myriophyllum aquaticum	River water	DO (0.1–5.0 mg/L), BOD (75%), pH (7.5–8.5), COD (67.4%), Tempt (15°C–19°C), TKN (88.3%), TP (93.6%).	30 days/ reactor tanks	Souza et al. (2013)
Pistia stratiotes Eichhornia crassipes Myriophyllum spicatum (Eurasian watermilfoil)	Contaminated river water	TN (77%), NH$_4^+$-N (93%), TP (88%), COD (61.70%); TN (87%), NH$_4^+$-N (96%), TP (84%), COD (68.21%); TN (77%), NH$_4^+$-N (93%), TP (80.55%), COD (62.55%).	6 months/ PVC water tanks	Lu et al. (2018)
Common reed, Manna grass, and Virginia mallow	Domestic wastewater	TSS (95%), BOD$_5$ (95%), TN (94%), COD (95%), and TP (95%).	5 years/ hybrid constructed wetland systems	Marzec et al. (2018)
Ceratophyllum demersum L. and *Stuckenia pectinata*	Fish pond	P (30.5–31.9 kg), N (448–842 kg).	3 months at 2-week interval	Petrů and Vymazal. (2018)
Alternanthera (Joyweed)	Domestic wastewater	Suspended solids (SS) (93%), TKN (45%), BOD (87%), COD (78.9%–83.9%), TP (36%), Cu (43%).	10 days/sheet flow root level (Shefrol bioreactor)	Abbasi et al. (2019)

(Continued)

TABLE 4.3 (*Continued*)

Macrophytes Used in Phytoremediation of Domestic and Agricultural Wastewater

Macrophytes	Domestic/ Agricultural Wastewater	Pollutant/Nutrient Reduction	Duration of Sample Study	Researchers
Nelumbo nucifera and *Nymphaea* (water lily)	Polluted surface water	BOD_5 (97.1 ± 0.8), turbidity (88.3 ± 0.3), COD (55 ± 1.6); BOD_5 (64.5 ± 0.3), turbidity (87.1 ± 0.6), COD (50.5 ± 2.3).	30 days/batch type	Abd Rasid et al. (2019)
Pistia stratiotes *Eichhornia crassipes* *Lemna minor* *Limnobium laevigatum*	Swine effluent	TP (23.8%), TN (63.2%); TP (15.4%), TN (60.2%); TP (36.2%), TN (60%); TP (28.2%), TN (48.8%).	3 months/ batch system	Sudiarto et al. (2019)
Salvinia natans	Domestic wastewater	PO_4-P (37%), TKN (85.2%), NO_2-N (40%), NH_4-N (79%), COD (95%), BOD_5 (96.9%).	8 months/ tanks	Laabassi and Boudehane (2019)
Typha orientalis *Sorghum arundinaceum*	Municipal wastewater	Turbidity (73.3%), phosphate (94.1%), Fe (98.1%), Na (74.7%), nitrate (17.8%), BOD (97.6%); turbidity (63.7%), nitrate (98%), BOD (96.1%), phosphate (95.2%), Fe (86.9%).	29 days/ sub-flow constructed wetlands	Oladejo et al. (2015)
Phragmites australis and Bacteria	Heavily contaminated Ravi River (Pakistan)	Total organic carbon (TOC) (86.6%), BOD_5 (83.3%), TP (2.4–0.5 mg/L), COD (85.9%), nitrate (33.3–1.2 mg/L), TN (37.5–2.1 mg/L), and trace metals: Ni (91.4%), Pb (36.14%), Fe (79.5%), Cr (85.1%), Mn (91.8%).	2 months/ batch system	Shahid et al. (2019)
Vetiveria zizanioides	Fish pond wastewater	NO_2 (27.51%), PO_4 (42.75%), NO_3 (25.5%), NH_3 (65.16%), NH_4 (30.17%).	6 weeks/ aquaculture system	Effendi et al. (2020)

TABLE 4.4
Macrophytes Used in Phytoremediation of Industrial Wastewater

Macrophytes	Industrial Wastewater	Accumulation/Nutrient Reduction	Duration	Researchers
Eichhornia crassipes *Pistia stratiotes*	Industrial wastewater	pH (7.78–7.35), BOD (16.4–8.1 mg/L), nitrite (183.4%), COD (122–33 mg/L), DO (7.69–5.6 mg/L), oil and grease (0.8–1.4 mg/L), Fe (139.4%); pH (7.8–7.3), BOD (17–8.3 mg/L), nitrite (0.7–0.2 mg/L), COD (126–5 mg/L), DO (7.5–5.3 mg/L), oil and grease (0.9–1.5 mg/L), Fe (137%).	2 months	Abdul Syukor et al. (2013)
Ceratophyllum demersum L., and *Potamogeton perfoliatus*	Danube river	Zn (104.2 µg/g), Pb (20.1 µg/g), Cd (3.5 µg/g), Cu (22.7 µg/g); Zn (58 µg/g), Pb (13.3 µg/g), Cd (1.9 µg/g), Cu (13.1 µg/g). Bioconcentration factors *(BCFs); Cd@ Zn > Pb > Cu*	19 months	Matache et al. (2013)
Alisma lanceolatum, Carex cuprina; Epilobium hirsutum; Juncus inflexus	Industrial effluents	Cd, Mn, Zn, Cr, Al, As, Ni, Cu, Pb, Fe, Sn, organic pollutants.	24 months/ constructed wetlands	Guittonny-Philippe et al. (2015)
Ipomoea aquatica	Synthetic wastewater	Pb uptake.		Chanu and Gupta, (2016)
Ceratophyllum demersum L.; Lemna gibba (Duckweed)	Synthetic aqueous wastewater	Ni (50%), Cd (82%) BCFs: Ni (104.16–200), Cd (75–707.92); Ni (50%), Cd (91%) BCFs: Ni (27.2–639), Cd (942.8–5093.3).	3 months/ (eighth day batch experiment)	Parnian et al. (2016)
Nasturtium officinale	Ammonium-rich wastewater	NH_3 (66%), COD (38%), Ca (23%), BOD (22%), total hardness (25%).	30 days	Musavi et al. (2016)
Azolla pinnata; S. molesta; Scirpus grossus	Pulp and paper mill effluent	COD (100%), colour (43.1%); COD (100%), colour (49.7%); COD (100%), colour (50.3%).	28 days	Ahmad et al. (2017)
Egeria densa Salvinia molesta	Batik wastewater	COD (95%), BOD (93%); COD (4338.5 mg/L), BOD (1693.1 mg/L).	17 days/ lab-scale reactor in batch systems	Tangahu and Putri (2017)

(Contiued)

TABLE 4.4 (*Continued*)
Macrophytes Used in Phytoremediation of Industrial Wastewater

Macrophytes	Industrial Wastewater	Accumulation/Nutrient Reduction	Duration	Researchers
Salvinia molesta	Heavy metal effluent	Hg (74%), Pb (85%).	10 days	Kumari et al. (2017)
Salvinia natans	Synthetic wastewater	Pb (9800 µg/g), Ni (42,363 µg/g), Cd ($\approx$6000 µg/g).	7 days	Leblebici et al. (2018)
S. cucullata; *Trapa natans*	Industrial wastewater; textile wastewater	TP (81.3%), COD (31.0%), NH$_3$-N (5.3%), DO (100%), nitrate (20%), BOD$_5$ (43%); BOD$_5$ (55%), COD (33.3%), DO (111.6%), nitrate (50%), NH$_3$-N (31.3%), TP (77.3%).	45 days/ batch cultures	Alam and Hoque (2017)
Typha angustifolia L.	Textile wastewater	Cd (28%), As (60%), Pb (45%), Cr (59%), BOD$_5$ (68%), COD (65%), TSS (35%), TDS (45%), colour (62%).	1 week/ constructed wetlands	Chandanshive et al. (2017)
Ludwigia stolonifera	Radioactive cesium (Cs) and cobalt (Co) wastewater	^{137}Cs (65%), ^{60}Co (95%).	20 days	Saleh et al. (2017)
Ipomoea aquatica	Palm oil mill wastewater	NH$_3$-N (82.7%), TSS (90%), nitrate (99%), TP (99%), COD (80%).	25 days/ bucket treatment system	Sa'at and Zaman (2017)
Typha angustifolia L.	Petrochemical wastewater	1,2-DCA (100%).	42 days/ subsurface wetlands	Al-Baldawi (2018)
Lemna minor	Institutional, textile, and distillery wastewater; industrial wastewater	pH (8–9), COD (92%), TDS (68%), BOD (92%), electrical conductivity (68%); pH (97.7%), COD (90.6%), BOD (85.6%), Zn (100.6%), Pb (107.5%), Cu (103.7%), TN (102.6%), TP (105%).	28 days 1 week	Amare et al. (2018) and Basiglini et al. (2018)
Vetiveria zizanioides	Tofu wastewater	pH (7.8), COD (71.78%), BOD (76%), TSS (75.28%).	15 days	Seroja et al. (2018)
Eichhornia crassipes Spirodela polyrhiza	Industrial wastewater	Cu (63%±42.4%), Ni (67±2.4%), Cd (76±3.4%), Fe (83%±4.4%), Zn (79%), Cr (66%±1.4%); As (37%±1.4%), Cd (48%), Cu (40%), Cr (50%), Fe (40%), Zn (60%).	15 days	Rai (2019)
Myriophyllum spicatum	Radioactive cesium (Cs) and cobalt (Co)	^{137}Cs (60%) and ^{60}Co (90%).	20 days contact time (40 hours)	Saleh et al. (2020)

TABLE 4.5

Taxonomy of *P. stratiotes*

Kingdom	Plantae
Subkingdom	Vascular plants
Subdivision	Seed plants
Division	Flowering plants
Class	Monocotyledons
Subclass	Arecidae
Order	Arales
Family	Araceae
Genus	Pistia
Species	*Pistia stratiotes* (water lettuce) (USDA, 2019)
Common names	Water lettuce, Nile cabbage, shellflower, water cabbage
Habitat	Wet and temperate habitat (lakes and rivers)

FIGURE 4.2   Description of *P. stratiotes* leaves.

their study. Additionally, Manjunath and Kousar (2016) employed the potentials of *P. stratiotes* in reducing the organic and inorganic pollutants in textile wastewater. The plant was acclimatized for 7 days and was left for 5 days in the wastewater effluent before commencement of the physicochemical analysis, whereas Aswathy (2017) constructed a small-scale wetland for remediation of kitchen wastewater (KWW) at 10 days detention time. Acclimatization of the plant in tap water before transplant into KWW was not reported. Furthermore, Yasar et al. (2018) carried out a comparative study between vertical and horizontal wetland systems cultivated with *P. stratiotes* and *Phragmites karka* plants for remediation of wastewater at a detention time of 12 hours for 3 days. The acclimatization period was not mentioned and the study was limited to physical, biological and chemical analysis of the wastewater samples. The relative growth rate (RGR) and doubling time of the two plants were not evaluated in the study. Additionally, Haidara et al. (2019) also used plastic containers in comparative study on varying weight of two aquatic plants for remediation of aquaculture wastewater at 21 days retention time using hydroponic systems. Physicochemical analyses (turbidity, pH, DO, temperature, BOD, NO_3, COD, PO_4^{3-}, and NH_4-N) were conducted every 7 days. Schwantes et al. (2019) used *P. stratiotes* in tertiary treatment of domestic wastewater at 42 days retention time. The plants were acclimatized for 15 days. The research was limited to sampling, characterization of the treated domestic wastewater, adaptability and inoculation of the *P. stratiotes*. Tabinda et al. (2019) investigated the performance of water hyacinth, alga (*Oedogonium* sp.) and water lettuce in phytoremediation of textile sewage containing dyes, Pb, Cu, Cd and Fe, for 7 days. The stoichiometric homeostatic index of *P. stratiotes* and the impacts of resource pulses were not studied, since resource pulse influences the metabolism, growth pattern, physiology and ecology of macrophyte species (Tan et al., 2019). Therefore, the metabolism, physiology and optimization of *P. stratiotes* in phytoremediation processes should be considered. However, this may vary according to the wastewater source, since each type of wastewater has a unique mixture of contaminants and nutrients. Furthermore, *P. stratiotes* has high biomass comprising of about 16.47% protein, 49.45% glucose, 3.56% fat and 17.81% raw fibres. As a result, *P. stratiotes* may be used as a bioenergy feedstock in the future (Kaur et al., 2018). Additionally, Pantawong et al. (2015) found that water lettuce produced a biogas yield of about 9667.33 mL at 45 days digestion time.

4.5.2 SALVINIA MOLESTA PLANTS

Salvinia molesta D.S. Mitchell commonly called Salvinia or giant salvinia is an aquatic plant of the Salviniaceae family. Salvinia was derived from Antonio M. Salvini, while molesta is a Greek term for irritation, troublesome or nuisance (Nelson, 2009). *S. molesta* has a high nutrient absorption capacity of 8 mg N/g of dry plant tissue per day (Nelson, 2009).

4.5.2.1 Distribution of *S. molesta*

S. molesta is an ancient plant that originates from Southeastern Brazil and has existed for over 70 years. It was initially discovered outside the territory of Sri Lanka in 1939. *S. molesta* has rapidly become a general weed issue in Sri Lanka that

TABLE 4.6

Taxonomy of *S. molesta*

Domain	Eukaryote
Kingdom	Plantae
Subkingdom	Tracheobionta
Phylum	Pteridophyta
Class	Liliopsida
Order	Hydropteridales
Family	Salviniaceae
Genera	Salvinia
Common names	Water spangles, African pyle, water moss, giant salvinia, Australian azolla, Kariba weed, water fern (McFarland et al., 2004; Ramachandra et al., 2017)
Species	*S. molesta, S. oblongifolia, S. natan, S. cucullata* (Nagalingum et al., 2008; Pérez-Consuegra et al., 2017), *S. herzogii, S. minima* Baker, *S. biloba Raddi, S. radula, S. auriculata, S. sprucei* (Avila & Funk, 2016), *S. hastata Desv, S. martynii Spruce* (McFarland et al., 2004)
Habitat	Lakes, dams and eutrophic environment (Ramachandra et al., 2017)

plagues rice paddies, decreases flow in water system (irrigation) and blocks transportation canal routes. Globally, *S. molesta* was considered one of the most harmful weeds (McFarland et al., 2004; Pérez-Consuegra et al., 2017). Furthermore, the plant was mainly found in tropical and subtropical regions of over 20 countries including Malaysia, Indonesia, Africa, Papua New Guinea, India, Singapore, Australia, Trinidad, Guyana, Zealand, Philippines, Fiji, Columbia and Madagascar (McFarland et al., 2004; Pérez-Consuegra et al., 2017).

4.5.2.2 Taxonomy of *S. molesta*

S. molesta belongs to a monogeneric family. The plant was originally thought to be a subspecies of *S. auriculata*. However, it was reclassified in 1972 based on evidence from fruiting bodies or male sporocarps (Nelson, 2009). There are 12 species of *S. molesta*, seven of which originate from the new world tropics. The detailed taxonomy of *S. molesta* is presented in Table 4.6.

4.5.2.3 Description of *S. molesta*

S. molesta is made up of fronds that occur in whorls of three leaves, with the first two leaves being round, buoyant and emergent, and the third leaf is submerged and separated to serve as the root. The buoyant fronds are arranged in oblong form, while the matured plants develop sacs in the form of eggs that contain infertile spores. Additionally, the top surface has cylindrical papillae, each of which is composed of four hairs at the distal end that are connected at their tips to create an inverted egg-beater shape. The cage-like structure of the end hairs is an active air trap that enables the plant to be resistant in water. The upper surface, papillae and end hairs of the plants are waterproof, while the lower layer of the leaves has an affinity for water. Overall, the plant exhibits considerable morphological variation in response

FIGURE 4.3 Description of *S. molesta* leaves.

to environmental factors such as available space, availability of nutrient and habitat. Additionally, it varies in size from less than 1.5 to 6cm broad leaves (GISD, 2005; Pieterse et al., 2003). Figure 4.3 presents the diagram of the *S. molesta* plants.

4.5.2.4 Growth of *S. molesta*

The growth of *S. molesta* depends on several factors including temperature, pH, nutrients, light and salinity (Owens & Smart, 2010). Additionally, the nutrient and temperature control the rate at which the plant develops (Madsen & Wersal, 2008). In proper conditions, *S. molesta* invasions are prevalent as the plants can double in biomass within a period of 2–3 days due to their asexual reproduction. The plants thrive at pH levels within 5–8. *S. molesta* can withstand temperatures between 5°C and 40°C (Madsen & Wersal, 2008). However, they perish within 2 hours of exposure to temperatures below −3°C or above 43°C. In addition to this, they are sensitive to changes in nutrients as they can directly absorb nutrients from wastewater. The discharge of agricultural effluents into rivers, lakes and streams increases the possibility of high yield of these plants as any increase in nutrient availability will significantly impact on the plant growth. This is due to phosphorous and nitrogen determining its growth rates and morphological characteristics. Likewise, the development of *S. molesta* is heavily restricted by salinity (Madsen & Wersal, 2008).

4.5.2.5   Nutrient Uptake by *S. molesta* Plants in Phytoremediation of Wastewater

S. molesta is a macrophyte species that has been used in wastewater phytoremediation because of their rapid growth and potentials to break down contaminants in wastewater (Hariyadi et al., 2013; Kumari et al., 2017). Pavithra and Kousar (2016) investigated the efficiency of *S. molesta* in reclamation of textile sewage. Ng et al. (2017) analysed the potentials of giant salvinia in remediation of wastewater for 16 days. Physicochemical parameters including turbidity, COD, NO_3, PO_4^{3-} and NH_4-N analysis were performed on the treated water. The outcome of the research showed that the excess nutrients present in the wastewater were lowered to 0.50 mg/L NO_3, 0.17 mg/L (95%) PO_4^3, and 2.62 mg/L NH_4-N. Macrophytes could store and absorb phosphorus, and their development rate is dependent on the phosphorous content in the wastewater. Abeywardhana et al. (2017) employed *S. molesta* and *L. gibba* in the remediation of industrial wastewater. WQ analysis was conducted on the wastewater at a detention time of 24 hours. The results demonstrated that up to 72.63% PO_4^3, 73.34% TN, 74.85% Pb, 66.39% Ni, 65.2% Fe, 69.81% Cu and 81.66% Cr reduction was obtained by *S. molesta* plants. Despite the duration of the sampling study not being mentioned by the authors, the results of the heavy metals and nutrient reduction efficiencies at 24 hours detention time revealed that *S. molesta* plants are effective in improving the contaminated water. Furthermore, reduction efficiency evaluation is critical, as it has been accepted as standard criteria for estimating the amount of nutrients removed during phytoremediation processes.

Additionally, Tangahu and Putri (2017) investigated the COD and BOD removal from batik wastewater using *E. densa* and *S. molesta* plants. It was observed that stress conditions were observed in the *S. molesta* plants on the fourth day of the cultivation period. This could be attributed to exposure of the *S. molesta* in the concentrated wastewater, implying that the wastewater was hazardous to the plant. In another study, Hanafiah et al. (2018) investigated the performance of water lettuce and giant salvinia in the treatment of wastewater. The findings established the effectiveness of *S. molesta* and *P. stratiotes* in polishing wastewater.

Furthermore, da Silva et al. (2018) evaluated the potentials of *S. molesta* in the remediation of polluted water. They also investigated the functions of peroxidase, ascorbate peroxidase, catalase and superoxide dismutase in *S. molesta* resistance to arsenite. The results of the study indicated that *S. molesta* had a higher proclivity for reducing arsenite's toxic effect in the polluted water. Sreekumar and John (2018) studied the potentials of giant salvinia in phytoremediation of pestilent water for bioaccumulation of Cd and Zn heavy metals. However, previous research mentioned failed to investigate the potentials of *S. molesta* in treatment of domestic wastewater, pharmaceutical, radioactive, nanoparticle and polymer-based materials using different weight (density) and retention times. Numerous reports have shown that the lipid content of giant salvinia varies between 15.72% and 19.97% depending on the pre-treatment technique used. Kaur et al. (2018) stated that the structure of unsaturated fats in giant salvinia is C14:0, C14:1, C16:0, C16:1, C18:0, C18:1, C20:1, C:20:4 and C22:0, comprising of fatty acids, monounsaturated (63.59%), polyunsaturated (0.73%) and saturated (33.16%) fats. Therefore, giant salvinia is a promising biomass for biofuel generation.

4.5.3 *EICHHORNIA CRASSIPES* PLANTS

Eichhornia crassipes (Mart.) Solms also called water hyacinth is a notorious macrophyte native to the Amazon Basin of South America. The plant belongs to the pickerelweed (Pontederiaceae) family and is considered one of the worst weed plants on the planet (Woldemichael et al., 2011). Water hyacinth has been used in the bioaccumulation of pollutants such as BOD, COD, phenols, pesticides, minerals, heavy metals and radioactive elements in phytoremediation processes (Khare & Lal, 2017).

4.5.3.1 Distribution of *E. crassipes*

E. crassipes plants are among the most distributed macrophytes in the world. They are usually found between 38°N and 38°S and were first identified as ornamental plant in the USA, South East Asia and South Africa in the 19th century. The weed plant was believed to spread easily in ponds and lakes of different parts of the world because of its attractive lavender flowers. Additionally, in 1988 *E. crassipes* was first discovered in Lake Kyoga (Uganda) in the zones circumscribing Lake Victoria. The plants were able to survive and grow in Lake Victoria because of the favourable habitat and absence of physical disturbance such as pests, fish and other biota (Gichuki et al., 2012).

4.5.3.2 Taxonomy of *E. crassipes*

E. crassipes belongs to the Pontederiaceae family, a taxonomically complicated family that was reclassified as a member of the Commelinales. *E. crassipes* is the only pantropical weed plant that has been identified. Except for *Eichhornia natans* (*P. Beauv.*), this family of mostly neotropical freshwater aquatics has eight additional genera and species that originates from South America (Coetzee et al., 2009). The detailed taxonomy of *E. crassipes* is presented in Table 4.7.

TABLE 4.7
Taxonomy of *E. crassipes*

Kingdom	Plantae
Subkingdom	Tracheobionta
Superdivision	Spermatophyta
Division	Magnoliophyta
Class	Liliopsida
Subclass	Liliidae
Order	Liliales
Family	Pontederiaceae
Genus	Eichhornia kunth
Species	*Eichhornia crassipes* (Mart.) Solms
Common name	Water hyacinth
Habitat	Surface of lakes, canals, ponds and rivers or mud of shallow waters (Ecoport, 2011; Pieterse, 1997). The most favourable conditions are K, N, P, low-salinity water, absence of natural predators, sunlight and optimum temperature of 25–30°C, while optimum pH is between 6 and 8.

FIGURE 4.4   Description of *E. crassipes* leaves.

4.5.3.3   Description of *E. crassipes*

E. crassipes plants are composed of stalked rosette leaves, feathery roots, slender root stocks and flowers arranged in clusters of leaf axils. The leaves are broadly circular, gently incurved, often undulate sides and have a diameter and height of 10–20 cm and 0.5–1 m, respectively. Additionally, the leaves are dark green in colour and are borne on inflated petioles resembling bladders (Ecoport, 2011). Leaf veins are numerous, thick, longitudinal and densely packed. As the axillary bud develops, leaves arise by rupturing a tabular leaf-like structure called a prophyll. Furthermore, the flowers have six petals with purplish-blue colour and yellow uppermost petals (Ecoport, 2011). The diagram of the plants displaying the flowers is presented in Figure 4.4.

4.5.3.4   Potentials of *E. crassipes* Plants
in Phytoremediation of Wastewater

E. crassipes has a high potential for phytoremediation of wastewater through bio-logical, chemical and physical processes (Mayo & Hanai, 2017). The plants have been used in phytoremediation of wastewater due to their ability to trap and degrade pollutants in wastewater. Nuraini and Felani (2015) studied the phytoremediation of liquid waste of tapioca industry using *E. crassipes* plants for 28 days. The study only considered seven WQ parameters. Fazal and Zhang (2015) investigated the potentials of *E. crassipes* in the treatment of industrial effluent. The findings demonstrated up to 92.5% turbidity, 83.7% COD, 91.8% TSS, 62.3% TDS, 80% TS, 71.6% NH_4-N, 67.5% NO_3, 90.2% PO_4^3, 97.5% Cd, 95.1% Ni, 99.9% Hg and 83.4% Pb reduction. Additionally, Qin et al. (2016) evaluated the phytoremediation of domestic wastewater using water hyacinth and water lettuce. The result of the study indicated that *E. crassipes* obtained a total reduction of 58.64% accumulating capacity for nitrogen. This result can be attributed to the leaf area and higher root activity (71.79–98.34 µg/g/h), root biomass (kg/m^2), net photosynthetic rate (20.28 µmol CO_2/m^2/S), active absorption area (0.31–0.36 m^2/g fresh weight) and large root surface (0.97–1.10 m^2/g fresh weight). Khare and Lal (2017) used a constructed wetland system of 5 ft × 4 ft × 2.5 ft dimension to investigate the potentials of *E. crassipes* in sewage treatment for 2 years. The physicochemical tests were restricted to turbidity TDS, BOD, DO, EC, COD, pH, K, P and nitrogen. Goswami and Das (2018) cultivated *E. crassipes* plants in synthetic wastewater containing different concentrations of Cu in Hoagland solution. The findings of the research indicated up to 55%–57% Cu reduction efficiency was obtained from the initial concentration of 5–10 mg/L Cu after 21 days. In addition, Nash et al. (2019) experimented the survival level and nutrient uptake of *E. crassipes* in Sago mill effluent (SME). The experiment was carried out in glass aquariums of 30 cm (length) × 30 cm (width) × 30 cm (depth) under controlled conditions for 30 days.

Furthermore, Safauldeen et al. (2019) used varying quantities of *E. crassipes* (8, 10 and 12 clumps) and retention times in phytoremediation of batik wastewater. Additionally, Sayago (2019) designed a sustainable method for phytoremediation of wastewater and generation of bioethanol using *E. crassipes*. In the study, the harvested *E. crassipes* biomass was loaded into bioreactors for bioethanol generation. The study proved the dual applications of *E. crassipes* for phytoremediation of wastewater and generation of biofuel using hydrolysis process. Parwin and Paul (2019) employed *E. crassipes* in the treatment of KWW for a 4-week period. Water quality parameters including turbidity, total hardness, NO_3-N, TDS, NH_4^+-N, K^+, PO_4^{3-}-P, EC, TOC, sulphate and bicarbonate were analysed. Although the study evaluated the reduction efficiency and bioconcentration factor, retention time and the estimated quantity of the plants used in the research were not mentioned. Besides, it is important to understand the nutrient assimilation, nutrient storage, nutrient uptake rates and growth rate of macrophytes.

Nevertheless, it was observed that recent research on phytoremediation methods of wastewater mentioned requires considerable effort and maintenance. Despite phytoremediation techniques improving the quality of the wastewater, the long-time monitoring and the vast area required for constructed wetland processes make the approach labour intensive. Thus, the use of phytoremediation techniques in the treatment of wastewater needs to be restricted. However, the importance of developing a simple, efficient and feasible technology such as hydroponic systems for phytoremediation of wastewater within a short period cannot be overemphasized.

4.6 CONCLUSION

This chapter presents the significance of macrophytes in the recovery and removal of pollutants from wastewater. The use of aquatic weed plants can be beneficial in these processes as they have tremendous capacity to absorb contaminants from water and they can be easily cultivated and managed. High-valued compounds used as feedstock for the generation of a wide scope of products can be derived from cultivation of aquatic plants, as they are a promising sustainable wellspring of supplements. Biomass obtained from phytoremediation can be utilized for biofuel and energy production, paper or charcoal generation, fertilizer and food supplement.

REFERENCES

Abbasi, S. A., Ponni, G., & Tauseef, S. M. (2019). Potential of joyweed Alternanthera sessilis for rapid treatment of domestic sewage in SHEFROL bioreactor. *International Journal of Phytoremediation, 21*(2), 1–10. https://doi.org/10.1080/15226514.2018.1488814

Abd Rasid, N. S., Naim, M. N., Che Man, H., Abu Bakar, N. F., & Mokhtar, M. N. (2019). Evaluation of surface water treated with lotus plant; Nelumbo nucifera. *Journal of Environmental Chemical Engineering, 7*(3). https://doi.org/10.1016/j.jece.2019.103048

Abdul Syukor, A. R., Zularisam, A.W., Ideris, Z., Mohd. Said, M. I., & Sulaiman, S. (2013). Treatment of industrial wastewater at Gebeng Area using Eichornia Crassipes Sp. (Water Hyacinth), Pistia Stratiotes Sp. (Water Lettuce) and Salvinia Molesta Sp. (Giant Salvinia). *Advances in Environmental Biology Journal, 7*(12), 3802–3807.

Abeywardhana, M. L. D. D., Bandara, N. J. G. J., & Rupasinghe, S. K. L. S. (2017). Waste management and pollution control. *Proceedings of the 22nd International Forestry and Environment Symposium 2017 of the Department of Forestry and Environmental Science, University of Sri Jayewardenepura, Sri Lanka, 233.*

Ahmad, J., Abdullah, S. R. S., Hassan, H. A., Rahman, R. A. A., & Idris, M. (2017). Saringan tumbuhan akuatik tropika tempatan untuk rawatan penyudahan sisa pulpa dan kertas. *Malaysian Journal of Analytical Sciences, 21*(1), 105–112. https://doi.org/10.17576/mjas-2017-2101-12

Akobundu, I. O., & Agyakwa, C. W. (1998). *A Handbook of African Weeds* (2nd ed.). International Institute of Tropical Agriculture, Ibadan.

Al-Baldawi, I. A. (2018). Removal of 1,2-Dichloroethane from real industrial wastewater using a sub-surface batch system with Typha angustifolia L. *Ecotoxicology and Environmental Safety, 147*(August 2017), 260–265. https://doi.org/10.1016/j.ecoenv.2017.08.022

Alam, A. K. M. R., & Hoque, S. (2017). Phytoremediation of industrial wastewater by culturing aquatic macrophytes, Trapa natans L. and Salvinia cucullata Roxb. *Jahangirnagar University Journal of Biological Sciences, 6*(2), 19–27.

Almuktar, S. A. A. A. N., Abed, S. N., & Scholz, M. (2018). Wetlands for wastewater treatment and subsequent recycling of treated effluent : A review. *Environmental Science and Pollution Research, 25*, 23595–23623.

Amare, E., Kebede, F., & Mulat, W. (2018). Wastewater treatment by Lemna minor and Azolla filiculoides in tropical semi-arid regions of Ethiopia. *Ecological Engineering, 120*, 464–473. https://doi.org/10.1016/j.ecoleng.2018.07.005

Anand, S., Bharti, S. K., Dviwedi, N., Barman, S. C., & Kumar, N. (2017). Macrophytes for the Reclamation of Degraded Waterbodies with Potential for Bioenergy Production. In K. Bauddh, B. Singh, & J. Korstad (Eds.), *Phytoremediation Potential of Bioenergy Plants* (pp. 1–472). Springer Nature, Singapore. https://doi.org/10.1007/978-981-10-3084-0

Aswathy, M. (2017). Wastewater treatment using constructed wetland with water lettuce (Eichornia Crasipies). *International Journal of Civil Engineering and Technology, 8*(8), 1413–1421.

Avila, F. A., & Funk, V. A. (2016). Líquenes a lythraceae Catálogo de plantas y líquenes. Brunelliaceae. In R. Bernal, S. R. Gradstein, M. Celis, & C. I. Orozco (Eds.), *Catálogo de plantas y líquenes de Colombia* (Primera ed, Issue April). Universidad Nacional de Colombia: Colombia.

Basiglini, E., Pintore, M., & Forni, C. (2018). Effects of treated industrial wastewaters and temperatures on growth and enzymatic activities of duckweed (Lemna minor L.). *Ecotoxicology and Environmental Safety*, 153(February), 54–59. https://doi.org/10.1016/j.ecoenv.2018.01.053

CABI. (2019). *Fallopia japonica*. In: Invasive Species Compendium. CAB International, Wallingford, UK. www.cabi.org/isc/datasheet/414196

Chandanshive, V. V., Rane, N. R., Tamboli, A. S., Gholave, A. R., Khandare, R. V, & Govindwar, S. P. (2017). Co-plantation of aquatic macrophytes Typha angustifolia and Paspalum scrobiculatum for effective treatment of textile industry effluent. *Journal of Hazardous Materials*, *338*, 47–56. https://doi.org/10.1016/j.jhazmat.2017.05.021

Chanu, L. B., & Gupta, A. (2016). Phytoremediation of lead using Ipomoea aquatica Forsk in hydroponic solution. *Chemosphere*, *156*, 407–411. https://doi.org/10.1016/j.chemosphere.2016.05.001

Coetzee, J., Hill, M., Julien, M., Center, T., & Cordo, H. (2009). Eichhornia crassipes (Mart.) Solms–Laub. (Pontederiaceae). In R. Muniappan, G. V. P. Reddy, & A. Raman (Eds.), *Biological Control of Tropical Weeds Using Arthropods* (Issue January, pp. 183–210). Cambridge University Press, Cambridge. https://doi.org/10.1017/CBO9780511576348.011

da Silva, A. A., de Oliveira, J. A., de Campos, F. V., Ribeiro, C., Farnese, F. dos S., & Costa, A. C. (2018). Phytoremediation potential of Salvinia molesta for arsenite contaminated water: role of antioxidant enzymes. *Theoretical and Experimental Plant Physiology*, *30*(4), 275–286. https://doi.org/10.1007/s40626-018-0121-6

Daverey, A., Pandey, D., Verma, P., Verma, S., Shah, V., Dutta, K., & Arunachalam, K. (2019). Recent advances in energy efficient biological treatment of municipal wastewater. *Bioresource Technology Reports*, 7(March). https://doi.org/10.1016/j.biteb.2019.100252

Dhir, B., Sharmila, P., & Saradhi, P. P. (2009). Potential of aquatic macrophytes for removing contaminants from the environment. *Critical Reviews in Environmental Science and Technology*, *39*(9), 754–781. https://doi.org/10.1080/10643380801977776

Dipu, S., Anju, A., Kumar, V., & Thanga, S. G. (2010). Phytoremediation of dairy effluent by constructed wetland technology using wetland macrophytes. *Global Journal of Environmental Research*, *4*(2), 90–100.

Ecoport. (2011). Ecoport. Ecoport Database.

Effendi, H., Utomo, B. A., & Pratiwi, N. T. M. (2020). Ammonia and orthophosphate removal of tilapia cultivation wastewater with Vetiveria zizanioides. *Journal of King Saud University - Science*, *32*(1), 207–212. https://doi.org/10.1016/j.jksus.2018.04.018

Ekperusi, A. O., Sikoki, F. D., & Nwachukwu, E. O. (2019). Application of common duckweed (Lemna minor) in phytoremediation of chemicals in the environment: State and future perspective. *Chemosphere*, *223*, 285–309. https://doi.org/10.1016/j.chemosphere.2019.02.025

Fazal, S., & Zhang, B. (2015). Biological treatment of combined industrial wastewater. *Ecological Engineering*, *84*, 551–558. https://doi.org/10.1016/j.ecoleng.2015.09.014

Flathman, P. E., & Lanza, G. R. (1998). Phytoremediation: Current views on an emerging green technology. *Journal of Soil Contamination*, *7*(4), 415–432. https://doi.org/10.1080/10588339891334438

George, G. T., & Gabriel, J. J. (2017). Phytoremediation of heavy metals from municipal waste water by Salvinia molesta Mitchell. *The Saudi Journal of Life Sciences*, *2*(3), 108–115. https://doi.org/10.21276/haya

Gichuki, J., Omondi, R., Boera, P., Okorut, T., Matano, A. S., Jembe, T., & Ofulla, A. (2012). Water hyacinth Eichhornia crassipes (Mart.) Solms-Laubach dynamics and succession in the Nyanza Gulf of Lake Victoria (East Africa): Implications for water quality and biodiversity conservation. *The Scientific World Journal, 2012*(106429), 1–10. https://doi.org/10.1100/2012/106429

GISD. (2005). *Species profile Salvinia molesta.* Global Invasive Species Database. http:www.iucngisd.org/gisd/species.php?sc=569

Goswami, S., & Das, S. (2018). Eichhornia crassipes mediated copper phytoremediation and its success using cat fish bioassay. *Chemosphere, 210,* 440–448. https://doi.org/10.1016/j.chemosphere.2018.07.044

Guittonny-Philippe, A., Petit, M. E., Masotti, V., Monnier, Y., Malleret, L., Coulomb, B., Combroux, I., Baumberger, T., Viglione, J., & Laffont-Schwob, I. (2015). Selection of wild macrophytes for use in constructed wetlands for phytoremediation of contaminant mixtures. *Journal of Environmental Management, 147,* 108–123. https://doi.org/10.1016/j.jenvman.2014.09.009

Haidara, A. M., Magami, I. M., & Sanda, A. (2018). Bioremediation of Aquacultural Effluents Using Hydrophytes. *Bioprocess Engineering, 2*(4), 33–37. https://doi.org/10.11648/j.be.20180204.11

Hanafiah, M. M., Mohamad, N. H. S., & Abd. Aziz, N. I. H. (2018). Salvinia molesta dan Pistia stratiotes sebagai Agen Fitoremediasi dalam Rawatan Air Sisa Kumbahan. *Sains Malaysiana, 47*(8), 1625–1634. https://doi.org/http://dx.doi.org/10.17576/jsm-2018-4708-01

Hanks, N. A., Caruso, J. A., & Zhang, P. (2015). Assessing Pistia stratiotes for phytoremediation of silver nanoparticles and Ag(I) contaminated waters. *Journal of Environmental Management, 164,* 41–45. https://doi.org/10.1016/j.jenvman.2015.08.026

Hariyadi, B., Yanuwiadi, B., & Polii, S. (2013). Phytoremediation of arsenic from geothermal power plant waste water using Monochoria vaginalis, Salvinia molesta and Colocasia esculenta. *International Journal of Biosciences (IJB), 3*(6), 104–111. https://doi.org/10.12692/ijb/3.6.104-111

Kaur, M., Kumar, M., Sachdeva, S., & Puri, S. K. (2018). Aquatic weeds as the next generation feedstock for sustainable bioenergy production. *Bioresource Technology, 251,* 390–402. https://doi.org/10.1016/j.biortech.2017.11.082

Khare, A., & Lal, E. P. (2017). Wastewater purification potential of Eichhornia crassipes (Water hyacinth). *International Journal of Current Microbiology and Applied Sciences, 6*(12), 3723–3731. https://doi.org/10.20546/ijcmas.2017.612.429

Kochi, L. Y., Freitas, P. L., Maranho, L. T., Juneau, P., & Gomes, M. P. (2020). Aquatic macrophytes in constructed wetlands: A fight against water pollution. *Sustainability (Switzerland), 12*(21), 1–21. https://doi.org/10.3390/su12219202

Kumari, S., Baidyanath, K., & Sheel, R. (2017). Biological control of heavy metal pollutants in water by Salvinia molesta. *International Journal of Current Microbiology and Applied Sciences, 6*(4), 2838–2843. https://doi.org/10.20546/ijcmas.2017.604.325

Kutty, S. R. M., Ngatenah, S. N. I., Mohamed, H. I., & Malakahmad, A. (2009). Nutrients removal from municipal wastewater treatment plant effluent using Eichhornia crassipes. *World Academy of Science, Engineering and Technology, 60,* 1115–1123.

Laabassi, A., & Boudehane, A. (2019). Wastewater treatment by floating macrophytes (Salvinia natans) under algerian semi-arid climate. *European Journal of Engineering and Natural Sciences, 3*(1), 103–110.

Labrada, R., & Fornasari, L. (2002). *Management of Problematic Aquatic Weeds in Africa. Food and Agriculture Organisation (FAO) Efforts and Achievement during the Period 1991–2001 of the United Nations.* FAO, Rome.

Leblebici, Z., Kar, M., & Yalçin, V. (2017). Comparative study of Cd, Pb, and Ni removal potential by Salvinia natans (L.) All. and Lemna minor L.: Interactions with growth parameters. Romanian Biotechnological Letters, x, 1–14.

Lima, L. K. S., Tosi, P. B., da Silva, M. G. C., & Vieira, M. G. A. (2013). Lead and chromium biosorption by pistia stratiotes biomass. Chemical Engineering Transactions, 32, 1045–1050. https://doi.org/10.3303/CET1332175

Lu, B., Xu, Z., Li, J., & Chai, X. (2018). Removal of water nutrients by different aquatic plant species: An alternative way to remediate polluted rural rivers. Ecological Engineering, 110, 18–26. https://doi.org/10.1016/j.ecoleng.2017.09.016

Lu, Q., He, Z. L., Graetz, D. A., Stoffella, P. J., & Yang, X. (2010). Phytoremediation to remove nutrients and improve eutrophic stormwaters using water lettuce (Pistia stratiotes L.). Environmental Science and Pollution Research, 17(1), 84–96. https://doi.org/10.1007/s11356-008-0094-0

Madsen, J. D., & Wersal, R. M. (2008). Growth regulation of Salvinia molesta by pH and available water column nutrients. Journal of Freshwater Ecology, 23(2), 305–313. https://doi.org/10.1080/02705060.2008.9664203

Magwaza, S. T., Magwaza, L. S., Odindo, A. O., & Mditshwa, A. (2020). Hydroponic technology as decentralised system for domestic wastewater treatment and vegetable production in urban agriculture: A review. Science of the Total Environment, 698, 134154. https://doi.org/10.1016/j.scitotenv.2019.134154

Mahmood, Q., Zheng, P., Siddiqi, M. R., Islam, E. U., Azim, M. R., & Hayat, Y. (2005). Anatomical studies on water hyacinth (Eichhornia crassipes (Mart.) Solms) under the influence of textile wastewater. Journal of Zhejiang University: Science, 6B(10), 991–998. https://doi.org/10.1631/jzus.2005.B0991

Manjunath, S., & Kousar, H. (2016). Phytoremediation of textile industry effluent using Pistia stratiotes. International Journal of Environmental Sciences, 5(2), 75–81.

Marzec, M., Jóźwiakowski, K., Dębska, A., Gizińska-Górna, M., Pytka-Woszczyło, A., Kowalczyk-Juśko, A., & Listosz, A. (2018). The efficiency and reliability of pollutant removal in a hybrid constructedwetland with common reed, Manna Grass, and Virginia Mallow. Water (Switzerland), 10(10). https://doi.org/10.3390/w10101445

Matache, M. L., Marin, C., Rozylowicz, L., & Tudorache, A. (2013). Plants accumulating heavy metals in the Danube River wetlands. Journal of Environmental Health Science and Engineering, 11(1). https://doi.org/10.1186/2052-336X-11-39

Mayo, A. W., & Hanai, E. E. (2017). Modeling phytoremediation of nitrogen-polluted water using water hyacinth (Eichhornia crassipes). Physics and Chemistry of the Earth, 100, 170–180. https://doi.org/10.1016/j.pce.2016.10.016

McFarland, D. G., Nelson, L. S., Grodowitz, M. J., Smart, R. M., & Owens, C. S. (2004). Salvinia molesta D.S. Mitchell (Giant Salivinia) in the United States: A review of species ecology and approaches to management. Aquatic Plant Control Research Program. http://el.erdc.army.mil/elpubs/pdf/srel04-2.pdf

Mishima, D., Kuniki, M., Sei, K., Soda, S., Ike, M., & Fujita, M. (2008). Ethanol production from candidate energy crops: Water hyacinth (Eichhornia crassipes) and water lettuce (Pistia stratiotes L.). Bioresource Technology, 99(7), 2495–2500. https://doi.org/10.1016/j.biortech.2007.04.056

Mukherjee, B., Majumdar, M., Gangopadhyay, A., Chakraborty, S., & Chaterjee, D. (2015). Phytoremediation of parboiled rice mill wastewater using water lettuce (Pistia Stratiotes). International Journal of Phytoremediation, 17(7), 651–656. https://doi.org/10.1080/15226514.2014.950415

Musavi, S. A., Karimi, N., & Sadeghi, S. (2016). Growth and Phytoremediation Potential of Watercress Nasturtium officinale R. Br. in Ammonium-rich Wastewater. Ambient Science, 3(2), 89–92. https://doi.org/10.21276/ambi.2016.03.2.ra08

Nagalingum, N. S., Nowak, M. D., & Pryer, K. M. (2008). Assessing phylogenetic relationships in extant heterosporous ferns (Salviniales), with a focus on Pilularia and Salvinia. *Botanical Journal of the Linnean Society, 157*, 673–685.

Nash, D. A. H., Abdullah, S. R. S., Hasan, H. A., Muhammad, N. F., Al-Baldawi, I. A., & Ismail, N. (2019). Phytoremediation of nutrients and organic carbon from sago mill effluent using water hyacinth (Eichhornia crassipes). *Journal of Engineering and Technological Sciences, 51*(4), 573–584. https://doi.org/10.5614/j.eng.technol.sci.2019.51.4.8

Nelson, L. S. (2009). Giant and Common Salvinia (pp. 157–164). www.aquatics.org/bmp-chapters/Chapter_15-09.pdf

Neuenschwander, P., Julien, M. H., Center, T. D., & Hill, M. P. (2009). Pistia stratiotes L. (Araceae). In R. Muniappan, G. V. P. Reddy, & A. Raman (Eds.), *Biological Control of Tropical Weeds Using Arthropods* (pp. 332–352). Cambridge University Press, Cambridge. https://doi.org/10.1017/CBO9780511576348.017

Ng, Y. S., Samsudin, N. I. S., & Chan, D. J. C. (2017). Phytoremediation capabilities of Spirodela polyrhiza and Salvinia molesta in fish farm wastewater: A preliminary study. *IOP Conference Series: Materials Science and Engineering, 206*(1). https://doi.org/10.1088/1757-899X/206/1/012084

Nivetha, C., Subraja, S., Sowmya, R., & Induja, N. M. (2016). Water lettuce for removal of nitrogen and phosphate from sewage. *IOSR Journal of Mechanical and Civil Engineering (IOSR-JMCE), 13*(2), 104–107. https://doi.org/10.9790/1684-13020198101

Nuraini, Y., & Felani, M. (2015). Phytoremediation of tapioca wastewater using water hyacinth plant (Eichhornia crassipes). *Journal of Degraded and Mining Lands Management, 2*(2), 295–302. https://doi.org/10.15243/jdmlm.2014.022.295

Oladejo, O. S., Ojo, O. M., & Akinpelu, O. I. (2015). *Wastewater Treatment Using Constructed Wetland With Water Lettuce (Pistia Stratiotes). International Journal of Chemical, Environmental and Biological Sciences, 3*(2), 119–124.

Owens, C. S., & Smart, R. M. (2010). *Effects of Nutrient, Salinity, and pH on Salvinia molesta (Mitchell) Growth. ERDC/TN APCRP-EA-23.U.S. Army Engineer Research and Development Centre, Vicksburg, MS.*

Pantawong, R., Chuanchai, A., Thipbunrat, Y., Unpaprom, P., & Ramaraj, R. (2015). Experimental investigation of biogas production from water lettuce, Pistia stratiotes L. *Emergent Life Sciences Research, 1*(2), 41–46.

Parnian, A., Chorom, M., Jaafarzadeh, N., & Dinarvand, M. (2016). Use of two aquatic macrophytes for the removal of heavy metals from synthetic medium. *Ecohydrology and Hydrobiology, 16*(3), 194–200. https://doi.org/10.1016/j.ecohyd.2016.07.001

Parwin, R., & Paul, K. K. (2019). Phytoremediation of kitchen wastewater using Eichhornia crassipes. *Journal of Environmental Engineering, 145*(6), 1–10. https://doi.org/10.1061/(ASCE)EE.1943-7870.0001520.

Patel, D. K., & Kanungo, V. K. (2012). Treatment of domestic wastewater by potential application of a submerged aquatic plant Hydrilla verticillata Casp. *Recent Research in Science and Technology, 4*(10), 56–61.

Pavithra, M., & Kousar, H. (2016). Potential of salvinia molesta for removal of sodium in textile wastewater. *Journal of Bioremediation & Biodegradation, 7*(5). https://doi.org/10.4172/2155-6199.1000364

Pérez-Consuegra, N., Cuervo-Gómez, A., Martínez, C., Montes, C., Herrera, F., & Jaramillo, C. (2017). Paleogene Salvinia (Salviniaceae) from Colombia and their paleobiogeographic implications. *Review of Palaeobotany and Palynology, 246*(June), 85–108. https://doi.org/10.1016/j.revpalbo.2017.06.003

Petrů, A., & Vymazal, J. (2018). Potential of submerged vegetation to remove nutrients from eutrophic fishponds. *Scientia Agriculturae Bohemica, 49*(4), 313–324. https://doi.org/10.2478/sab-2018-0038

Pieterse, A H. (1997). Eichhornia crassipes (Martius) Solms. In H. I. Faridah & L. J. G. Van der Maesen (Eds.), *Record from Proseabase*. PROSEA (Plant Resources of South-East Asia) Foundation, Bogor, Indonesia.

Pieterse, A. H., Kettunen, M., Diouf, S., Ndao, I., Sarr, K., Tarvainen, A., Kloff, S., & Hellsten, S. (2003). Effective biological control of Salvinia molesta in the Senegal River by Means of the weevil Cyrtobagous salviniae. *AMBIO: A Journal of the Human Environment*, *32*(7), 458–462. https://doi.org/10.1579/0044-7447-32.7.458

Qin, H., Zhang, Z., Liu, M., Liu, H., Wang, Y., Wen, X., Zhang, Y., & Yan, S. (2016). Site test of phytoremediation of an open pond contaminated with domestic sewage using water hyacinth and water lettuce. *Ecological Engineering*, *95*, 753–762. https://doi.org/10.1016/j.ecoleng.2016.07.022

Rai, P. K. (2019). Heavy metals/metalloids remediation from wastewater using free floating macrophytes of a natural wetland. *Environmental Technology and Innovation*, *15*, 1–8. https://doi.org/10.1016/j.eti.2019.100393

Raju, A. R., Anitha, C. T., Sidhimol, P. D., & Rosna, K. J. (2010). Phytoremediation of Domestic Wastewater by Using a Free Floating Aquatic Angiosperm, Lemna minor. *Nature Environment and Pollution Technology*, *9*(1), 83–88. www.neptjournal.com

Ramachandra, T. V., Bhat, S. P., & Vinay, S. (2017). Constructed wetlands for tertiary treatment of wastewater (ENVIS Technical Report 124) (Issue October). http://ces.iisc.ernet.in/energy/, http://ces.iisc.ernet.in/biodiversity

Rivers, L. (2002). Water Lettuce (Pistia stratiotes). University of Florida and Sea Grant, Gainesville.

Sa'at, S. K. M., & Zaman, N. Q. (2017). Suitability of Ipomoea aquatica for the treatment of effluent from palm oil mill. *Journal of Built Environment, Technology and Engineering*, *2*(May), 39–44.

Safauldeen, S. H., Hasan, H. A., & Abdullah, S. R. S. (2019). Phytoremediation efficiency of water hyacinth for batik textile effluent treatment. *Journal of Ecological Engineering*, *20*(9), 177–187. https://doi.org/10.12911/22998993/112492

Saleh, H. M., Bayoumi, T. A., Mahmoud, H. H., & Aglan, R. F. (2017). Uptake of cesium and cobalt radionuclides from simulated radioactive wastewater by Ludwigia stolonifera aquatic plant. *Nuclear Engineering and Design*, *315*, 194–199. https://doi.org/10.1016/j.nucengdes.2017.02.018

Saleh, H. M., Moussa, H. R., Mahmoud, H. H., El-Saied, F. A., Dawoud, M., & Abdel Wahed, R. S. (2020). Potential of the submerged plant Myriophyllum spicatum for treatment of aquatic environments contaminated with stable or radioactive cobalt and cesium. *Progress in Nuclear Energy*, *118*. https://doi.org/10.1016/j.pnucene.2019.103147

Sayago, U. F. C. (2019). Design of a sustainable development process between phytoremediation and production of bioethanol with Eichhornia crassipes. *Environmental Monitoring and Assesment*, *191*(221), 1–8. https://doi.org/10.1007/s10661-019-7328-0

Schwantes, D., Gonçalves, A. C., Schiller, A. da P., Manfrin, J., Campagnolo, M. A., & Somavilla, E. (2019). Pistia stratiotes in the phytoremediation and post-treatment of domestic sewage. *International Journal of Phytoremediation*, *21*(7), 714–723. https://doi.org/10.1080/15226514.2018.1556591

Seroja, R., Effendi, H., & Hariyadi, S. (2018). Tofu wastewater treatment using vetiver grass (Vetiveria zizanioides) and zeliac. *Applied Water Science*, *8*(1), 1–6. https://doi.org/10.1007/s13201-018-0640-y

Shahid, M. J., Arslan, M., Siddique, M., Ali, S., Tahseen, R., & Afzal, M. (2019). Potentialities of floating wetlands for the treatment of polluted water of river Ravi, Pakistan. *Ecological Engineering*, *133*, 167–176. https://doi.org/10.1016/j.ecoleng.2019.04.022

Sharma, P., & Pandey, S. (2014). Status of phytoremediation in world scenario. *International Journal of Environmental Bioremediation and Biodegradation*, *2*(4), 178–191. https://doi.org/10.12691/ijebb-2-4-5

Singh, D., Tiwari, A., & Gupta, R. (2012). Phytoremediation of lead from wastewater using aquatic plants. *Journal of Agricultural Technology, 8*(81), 1–11.

Souza, F. A., Dziedzic, M., Cubas, S. A., & Maranho, L. T. (2013). Restoration of polluted waters by phytoremediation using Myriophyllum aquaticum (Vell.) Verdc., Haloragaceae. *Journal of Environmental Management, 120,* 5–9. https://doi.org/10.1016/j.jenvman.2013.01.029

Sreekumar, A., & John, J. (2018). Removal of pollutants from pestilent water using selected hydrophytes. *Trends in Biosciences, 11*(7), 1618–1621.

Sudiarto, S. I. A., Renggaman, A., & Choi, H. L. (2019). Floating aquatic plants for total nitrogen and phosphorus removal from treated swine wastewater and their biomass characteristics. *Journal of Environmental Management, 231,* 763–769. https://doi.org/10.1016/j.jenvman.2018.10.070

Tabinda, A. B., Arif, R. A., Yasar, A., Baqir, M., Rasheed, R., Mahmood, A., & Iqbal, A. (2019). Treatment of textile effluents with Pistia stratiotes, Eichhornia crassipes and Oedogonium sp. *International Journal of Phytoremediation, 21*(10), 939–943. https://doi.org/10.1080/15226514.2019.1577354

Tangahu, B. V., & Putri, A. P. (2017). The degradation of BOD and COD of batik industry wastewater using Egeria densa and Salvinia molesta. *Jurnal Sains & Teknologi Lingkungan, 9*(2), 82–91. https://doi.org/10.20885/jstl.vol9.iss2.art2

Thongtha, S., Teamkao, P., Boonapatcharoen, N., Tripetchkul, S., Techkarnjararuk, S., & Thiravetyan, P. (2014). Phosphorus removal from domestic wastewater by Nelumbo nucifera Gaertn. and Cyperus alternifolius L. *Journal of Environmental Management, 137,* 54–60. https://doi.org/10.1016/j.jenvman.2014.02.003

Ting, W. H. T., Tan, I. A. W., Salleh, S. F., & Wahab, N. A. (2018). Application of water hyacinth (Eichhornia crassipes) for phytoremediation of ammoniacal nitrogen: A review. *Journal of Water Process Engineering, 22*(October), 239–249. https://doi.org/10.1016/j.jwpe.2018.02.011

USDA. (2019). Department of Agriculture, the PLANT Database, National Plant Data Team. NRCS, United States Greensboro, NC 27401-4901 USA (Accessed 21 October 2019).

Wickramasinghe, S., & Jayawardana, C. K. (2018). Potential of aquatic macrophytes Eichhornia crassipes, Pistia stratiotes and Salvinia molesta in phytoremediation of textile wastewater. *Journal of Water Security, 4,* 1–8. https://doi.org/10.15544/jws.2018.001

Woldemichael, D., Zewge, F., & Leta, S. (2011). Potential of water hyacinth. *Ethiopian Journal of Science, 34*(1), 49–62.

Yasar, A., Zaheer, A., Tabinda, A. B., Khan, M., Mahfooz, Y., Rani, S., & Siddiqua, A. (2018). Comparison of reed and water lettuce in constructed wetlands for wastewater treatment. *Water Environment Research, 90*(2), 129–135. https://doi.org/10.2175/106143017x14902968254728

5 Phytoremediation of Wastewater in Hydroponic Systems

5.1 INTRODUCTION

Hydroponic system can be described as a technique of plant cultivation in the absence of soil (Aires, 2018). Hydroponic ponds can be used to grow aquatic plants in nutrient-rich effluent. This method is regarded as an essential stage in wastewater phytoremediation, in which plant roots break down the excess nutrients present in wastewater (Norström et al., 2004). Therefore, the application of hydroponic systems would promote sustainable method for effective wastewater management (Baddadi et al., 2019). Hydroponic systems are classified based on the type of nutrient feeding (Hussain et al., 2014). Furthermore, they may be divided into two broad groups: nutrient film and hydroponic systems that do not require a substrate (Cooper, 1979). The substrate-based culture methods enable root anchoring and serve as a medium for microbe attachment, as well as water-nutritional flywheels (Magwaza et al., 2020). Besides, factors such as sunlight, plant roots, wastewater composition and growth of microorganism play a significant role for effective treatment of wastewater and biomass production in hydroponic systems. The plant root system also promotes the growth of microorganisms by generating metabolites that act as food for the bacteria, thereby enhancing the rate of nitrification and denitrification process in biological wastewater treatment (Magwaza et al., 2020). Additionally, hydroponic cultivation using wastewater would promote low-cost cultivation of useful crops and biomass, ensuring food security and creating income for underprivileged areas (Snow et al., 2008).

5.2 OVERVIEW OF HYDROPONIC SYSTEMS IN WASTEWATER TREATMENT

Hydroponics is derived from the Greek words hydro (water) and ponos (labour). In most cases, the hydroponic system is usually used for the tertiary stage treatment of wastewater (Yang et al., 2015). The construction of hydroponic systems for wastewater requires a multidisciplinary approach that incorporates design principles from the fields of civil engineering, biology, chemistry and biochemistry. Additionally, a greenhouse is introduced in the design to enhance the biological treatment system. The main purpose of the greenhouse is to provide climatic conditions that favour plant growth by protecting plants from adverse weather conditions, including temperatures, excessive precipitation and severe winds. The primary goal of the greenhouse

DOI: 10.1201/9781003359586-5

is to create eco-friendly conditions that would protect cultivated plants from adverse weather elements including wind, temperature and precipitation. Thus, it is best to supplement sunlight, energy and CO_2 to increase yield, particularly during winter season. However, it was discovered that a number of design factors connected to the design of hydroponics for wastewater treatment impede the construction standards (Magwaza et al., 2020). Additionally, wastewater treatment using hydroponic systems involves biological, chemical and physical processes with interaction between plant roots, microorganisms and nutrients. These systems are similar to constructed wetlands (Vipat et al., 2008). Furthermore, hydroponic systems are more sustainable, economical and environmentally friendly for use in wastewater remediation than constructed wetlands. Unlike the hydroponic systems, constructed wetland treatment systems utilize plants with minimal economic value to break down contaminants from wastewater (Vaillant et al., 2003). Moreover, hydroponic systems are easy to maintain, require less energy and can be implemented onsite within a small space (Abe et al., 2010).

5.3   NUTRIENT RECOVERY BY AQUATIC PLANTS

Tertiary treatment measures may be utilized to polish treated wastewater that has passed through the primary and secondary stages of conventional treatment methods. Tertiary treatment of effluents includes a progression of extra strides following secondary treatment to further reduce the amount of phosphorous, nitrogen, metals, organics, pathogens and turbidity existing in wastewater. In summary, tertiary sewage is treated when the wastewater is reused for drinking, agricultural and leisure purposes (Gerba & Pepper, 2015). Besides, plants play a significant role in supplying oxygen through its roots, which aid the processes that lead to degradation of residual pollutants (Stewart et al., 2008). These processes include absorption of nutrients and heavy metals, denitrification and nitrification, and other biological processes in the treatment systems. Therefore, absorption of inorganic and organic pollutants by aquatic plants (macrophytes) through volatilization is the essential process of nutrient uptake and microbial transformation in wastewater (Stewart et al., 2008). Thus, emergent, submergent and floating macrophytes have been used in reducing pollutants from different types of wastewater (Muradov et al., 2014). This is due to the ability of the macrophytes to grow in the wastewater while absorbing, accumulating and storing contaminants or nutrients in their roots and shoots through hydroponic cultivation of plants and transplanting into polluted waters. Plant roots contribute greatly to water component uptake through multiple factors (Sharma & Gaur, 1995). Macrophytes accumulate substantial amount of pollutants: phosphates, nitrates, ammonia, heavy metals, radioactive compounds and other toxic compounds at low cost with low energy using eco-friendly approaches with limited secondary waste disposal. They also produce large quantities of biomass roots through the use of natural renewable sources. Recently, the discovery of aquatic plants as phytoremediators and feedstock for the production of biofuels has led to its usage for gathering nutrients from the roots to the rhizosphere and to filter suspended solids. Several researchers maintain that macrophytes can discharge oxygen from roots to encompassing rhizosphere while providing aerobic conditions for nitrification occurrence.

5.4 CASE STUDY: DEMONSTRATION OF AQUATIC PLANTS CULTIVATION IN WASTEWATER

The cultivation of aquatic plants for nutrient uptake using hydroponic systems involves different stages. The stages are described in the subsequent sections.

5.4.1 CULTIVATION AREA

The cultivation of the plants was performed using a lab-scale hydroponic system. The cultivation area was situated in Kajang (2°58′04″ N, 101°43′55″ E), Selangor, Malaysia. The yearly average temperature of the environment is 27°C. Additionally, the sewage treatment plant was separated into different chambers, where wastewater is treated at the primary and secondary stages before being released into natural environment. Figures 5.1 shows the location of the research site (Adapted from Google Earth).

5.4.2 RESEARCH SETUP

Four rectangular transparent hydroponic ponds measuring 670 mm (length) × 420 mm (width) × 220 mm (depth) were built for the cultivation of the macrophytes. A thick glass of 6 mm was placed in the centre of the pond to act as tray support and tank divider, segregating the influent and effluent flow of the water sample. Additionally, the digital timer and the submersible pump control the inlet and outlet water pipes. The control timer was connected to a power source located 40 m away from the hydroponic system. It regulates the inflow of the influent and outflow of the effluent water in the hydroponic system at the fixed retention time. The submersible pump

FIGURE 5.1 Map of West Malaysia and the research site. Adapted from Google Earth.

(a)

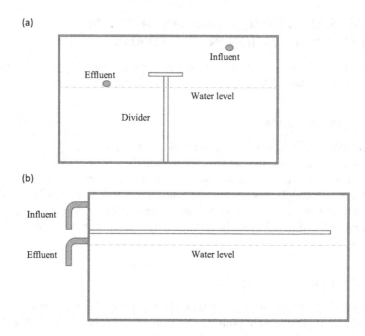

(b)

FIGURE 5.2   (a) Front view of the hydroponic tank; (b) side view of the hydroponic tank.

was installed at the discharge point of the secondary wastewater. Additionally, a detachable tray sheet perforated with 20 holes of 10–15 mm diameter was used to hold the plants vertically in place. Figure 5.2 illustrates the dimensions of the constructed hydroponic system. Furthermore, the constructed hydroponic ponds served as a shallow pond system for the cultivation of the aquatic plants and phytoremediation processes of the domestic wastewater.

5.4.3 SOURCE OF THE PLANT SAMPLES

P. stratiotes and *E. crassipes* plants were obtained within the premises of the research site, while *S. molesta* plants was collected from MRB nursery, Kajang, Selangor. The plants were rinsed with water to remove unwanted particles before weighing the fresh plants into each of the constructed treatment systems containing the influent water samples.

5.4.4 METHOD OF PLANT CULTIVATION

The selected plants were cultivated according to the methods reported by Kutty et al. (2009) and Rezania et al. (2015), but with adjustments in the acclimatization period and retention time. The plants were cultivated in three stages. In the first stage, 100 g each of the fresh *P. stratiotes*, *S. molesta* and *E. crassipes* were cultivated in separate hydroponic tanks containing the influent water samples from the secondary treated water. The plants were allowed to acclimatize in the systems for 5 days prior to the sampling study.

FIGURE 5.3 First-stage cultivation of *P. stratiotes*, *S. molesta* and *E. crassipes*.

FIGURE 5.4 Second-stage cultivation of different weights of *S. molesta* plants.

In the second stage, 70, 140 and 280 g of the fresh *S. molesta* plants were culti-vated and acclimatized in three constructed treatment systems containing the influ-ent water samples. The different retention time and weight of the plant was selected in order to optimize the operational process and conditions for the wastewater treat-ment and harvesting of the plants for feedstock applications (Figures 5.3 and 5.4).

5.4.5 Method of Water Sample Collection

Sampling was conducted every 2 days at 6-, 12- and 24-hour detention times for 14 days. The influent samples were collected directly from the discharge point of the secondary treatment, while the effluent samples were collected from each of the hydroponic tanks. The influent and effluent samples were collected aseptically in 250 mL bottles. The collected water samples were transferred to the environmental laboratory for water quality tests. In the second stage of the cultivation process, the influent and effluent water samples were collected every 2 days at 24 hours retention time. In some cases, the water samples were preserved in a refrigerator before labora-tory analysis.

5.5 CASE STUDY: RELATIVE GROWTH RATE (RGR) OF AQUATIC PLANTS IN PHYTOREMEDIATION SYSTEMS

Relative growth rate (RGR) is a parameter used in estimating plant growth. In this case study, RGR for *E. crassipes*, *P. stratiotes* and *S. molesta* at 24 hours retention time was evaluated. The harvesting of the plants was carried out once a week from the 2 weeks cultivation of the plants in hydroponic systems. The results were calculated using the equation below (Valipour et al., 2015). The graph of the RGR against the cultivated aquatic plants is presented in Figure 5.5.

$$\text{Relative growth rate } (\%) \ = \frac{\ln Q_2 - \ln Q_1}{r_2 - r_1} \times 100$$

where $Q1$ and $Q2$ are the dry weight of the plants at times $r1$ and $r2$, respectively.

According to Figure 5.5, it can be seen that the test plants grow efficiently from day 0 to day 14. The pattern of the RGR showed that the plants tripled in weight within the 14 days of the sampling study. *E. crassipes* recorded the highest growth with 2.5 ± 0.03 g/g/d, followed by *S. molesta* with 1.33 ± 0.05 g/g/d and *P. stratiotes* with 0.92 ± 0.27 g/g/d at the end of the sampling period. The high density observed by *E. crassipes* could be attributed to the long roots. ANOVA tests revealed a statistical difference between the growth rate of the individual test plants. Similarly, the high growth observed in the three plants was ascribed to the longer retention of 24 hours, which provided the plants with enough time to use the available nutrients present in the wastewater as food for their growth. Additionally, this finding revealed that the plants are capable of absorbing phosphates, nitrate and ammonia as nourishment.

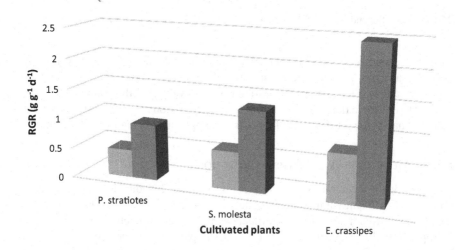

FIGURE 5.5 Graph of RGR (g/g/d) against cultivated plants.

The presence of nutrients accelerated the growth of the plants, making them more effective in reducing the contaminants present in the influent samples. Similarly, this result supports the findings of Ansari et al. (2017), who stated that wastewater is a rich source of nutrients, water and energy, which encourages plants to grow, reproduce and survive when employed in phytoremediation of wastewater processes. Furthermore, nitrogen and phosphorus are macronutrients that are required by all plants. Nitrogen is an essential component of all living creatures since it is found in DNA, protein, amino acids, ribonucleic acid (RNA) and other cell components (Adhikari et al., 2020). The residual nutrients present after the primary and secondary treatment of the raw domestic wastewater were the primary source of phosphates, ammonia nitrogen and nitrate in the treatment systems.

5.6 MANAGEMENT OF THE HARVESTED AQUATIC PLANTS

Aquatic weed plants utilized in wastewater treatment have high growth rates. Aquatic plants that exhibit this characteristic are considered invasive, as they are capable of rapid colonization (Fletcher et al., 2020). Hence, there exist a critical link between the application of macrophytes in wastewater treatment and approaches for management of invasive plants (Fletcher et al., 2020). However, it is more appropriate to incorporate these plants as part of an integrated management strategy that successfully curtails the spread of the plants and remediation of polluted water (Yan et al., 2016). Additionally, when considering wastewater treatment using aquatic plants, environmental managers should conduct preliminary study to determine the optimum balance of the nutrient input/output and plant removal capacity, as well as other upstream best practices. Manual technique (hand removal method) of aquatic harvesting plants is more suitable for lab-scale hydroponic system. Moreover, the hand removal method is tedious and time consuming, but it allows careful removal of selected unwanted weed (Quilliam et al., 2015).

5.7 CONCLUSION

Macrophytes are important in biological wastewater treatment due to their ability in eliminating contaminants depending on the retention time, nature of pollutants, ambient conditions and plant features. Emergent, submergent and floating macrophytes have been used in reducing pollutants from different wastewater sources. This is due to the ability of the macrophytes to grow in the wastewater while absorbing, accumulating and storing contaminants or nutrients in their roots and shoots through hydroponic cultivation of plants and transplanting into polluted waters. Furthermore, the outcome of the RGR demonstrated that the macrophytes grow efficiently from day 0 to day 14. Similarly, the trend of the RGR indicated that the plants tripled in weight within the 14 days of the sampling study. The high density observed by *E. crassipes* could be attributed to the long roots. In addition, the high growth observed in the three plants was ascribed to the longer retention of 24 hours, which provided the plants with enough time to use the available nutrients present in the wastewater as food for their growth. Additionally, this finding revealed that the plants are capable of absorbing phosphates, nitrate and ammonia as nourishment.

REFERENCES

Abe, K., Kato, K., & Ozaki, Y. (2010). Vegetation-based wastewater treatment technologies for rural areas in Japan. *Japan Agricultural Research Quarterly, 44*(3), 231–242. https://doi. org/10.6090/jarq.44.231

Adhikari, R., Rauniyar, S., Pokhrel, N., Wagle, A., Komai, T., & Paudel, S. R. (2020). Nitrogen recovery via aquaponics in Nepal: Current status, prospects, and challenges. *SN Applied Sciences, 2*(1192), 1–15. https://doi.org/10.1007/s42452-020-2996-5

Aires, A. (2018). Hydroponic production systems: Impact on nutritional status and bioactive compounds of fresh vegetables. *Vegetables - Importance of Quality Vegetables to Human Health.* https://doi.org/10.5772/intechopen.73011

Ansari, A. J., Hai, F. I., Price, W. E., Drewes, J. E., & Nghiem, L. D. (2017). Forward osmosis as a platform for resource recovery from municipal wastewater - A critical assessment of the literature. *Journal of Membrane Science, 529*(July 2016), 195–206. https://doi. org/10.1016/j.memsci.2017.01.054

Baddadi, S., Bouadila, S., Ghorbel, W., & Guizani, A. A. (2019). Autonomous greenhouse microclimate through hydroponic design and refurbished thermal energy by phase change material. *Journal of Cleaner Production, 211,* 360–379. https://doi.org/10.1016/j. jclepro.2018.11.192

Cooper, A. (1979). *The ABC of NFT: Nutrient Film Technique.* Grower Book, London.

Fletcher, J., Willby, N., Oliver, D. M., & Quilliam, R. S. (2020). Phytoremediation Using Aquatic Plants. In B. Shmaefsky (ed.), Issue April, *Phytoremediation: In-Situ Applications.* Springer International Publishing, Cham. https://doi.org/10.1007/978-3-030-00099-8

Gerba, C. P., & Pepper, I. L. (2015). Municipal Wastewater Treatment. In I. L. Pepper, C. P. Gerba, & T. J. Gentry (Eds.), *Environmental Microbiology* (3rd ed.). Elsevier Inc, Cambridge, United States. https://doi.org/10.1016/B978-0-12-394626-3.00025-9

Hussain, A., Iqbal, K., Aziem, S., Mahato, P., & Negi, A. K. (2014). A review on the science of growing crops without soil (soilless culture) – A novel alternative for growing crops. *International Journal of Agriculture and Crop Science, 7*(11), 833–842.

Kutty, S. R. M., Ngatenah, S. N. I., Mohamed, H. I., & Malakahmad, A. (2009). Nutrients removal from municipal wastewater treatment plant effluent using Eichhornia crassipes. *World Academy of Science, Engineering and Technology, 60,* 1115–1123.

Magwaza, S. T., Magwaza, L. S., Odindo, A. O., & Mditshwa, A. (2020). Hydroponic technology as decentralised system for domestic wastewater treatment and vegetable production in urban agriculture: A review. *Science of the Total Environment, 698,* 134154. https:// doi.org/10.1016/j.scitotenv.2019.134154

Muradov, N., Taha, M., Miranda, A. F., Kadali, K., Gujar, A., Rochfort, S., Stevenson, T., Ball, A. S., & Mouradov, A. (2014). Dual application of duckweed and azolla plants for wastewater treatment and renewable fuels and petrochemicals production. *Biotechnology for Biofuels, 7*(1), 1–17. https://doi.org/10.1186/1754-6834-7-30

Norström, A., Larsdotter, K., Gumaelius, L., La Cour Jansen, J., & Dalhammar, G. (2004). A small scale hydroponics wastewater treatment system under Swedish conditions. *Water Science and Technology, 48*(11–12), 161–167. https://doi.org/10.2166/wst.2004.0830

Quilliam, R. S., Niekerk, M. A. Van, Chadwick, D. R., Cross, P., Hanley, N., Jones, D. L., Vinten, A. J. A., Willby, N., & Oliver, D. M. (2015). Can macrophyte harvesting from eutrophic water close the loop on nutrient loss from agricultural land? *Journal of Environmental Management, 152,* 210–217. https://doi.org/10.1016/j.jenvman.2015.01.046

Rezania, S., Din, M., Taib, S., Dahalan, F., Songip, A., Singh, L., & Hesam, K. (2015). The efficient role of aquatic plant (water hyacinth) in treating domestic wastewater in continuous system. *International Journal of Phytoremediation.* https://doi.org/10.1080/152 26514.2015.1130018

Sharma, S. S., & Gaur, J. P. (1995). Potential of Lemna polyrrhiza for removal of heavy metals. *Ecological Engineering*, *4*(1), 37–43. https://doi.org/10.1016/0925-8574(94)00047-9

Snow, A. M., Ghaly, A. E., & Snow, A. (2008). A comparative assessment of hydroponically grown cereal crops for the purification of aquaculture wastewater and the production of fish feed. *American Journal of Agricultural and Biological Science*, *3*(1), 364–378. https://doi.org/10.3844/ajabssp.2008.364.378

Stewart, F. M., Mulholland, T., Cunningham, A. B., Kania, B. G., & Osterlund, M. T. (2008). Floating islands as an alternative to constructed wetlands for treatment of excess nutrients from agricultural and municipal wastes – Results of laboratory-scale tests. *Land Contamination and Reclamation*, *16*(1), 25–33.

Vaillant, N., Monnet, F., Sallanon, H., Coudret, A., & Hitmi, A. (2003). Treatment of domestic wastewater by an hydroponic NFT system. *Chemosphere*, *50*(1), 121–129. https://doi.org/10.1016/S0045-6535(02)00371-5

Valipour, A., Raman, V. K., & Ahn, Y. H. (2015). Effectiveness of domestic wastewater treatment using a Bio-hedge water hyacinth wetland system. *Water (Switzerland)*, *7*(1), 329–347. https://doi.org/10.3390/w7010329

Vipat, V., Singh, U. R., & Billore, S. K. (2008). *Efficacy of Rootzone Technology for Treatment of Domestic Wastewater: Field Scale Study of a Pilot Project in Bhopal* (MP), India. January 2008, 995–1003.

Yan, S., Song, W., Guo, J., Yan, S., Song, W., & Guo, J. (2016). Advances in management and utilization of invasive water hyacinth (Eichhornia crassipes) in aquatic ecosystems – A review. *Critical Reviews in Biotechnology*, *8551*(February), 1–11. https://doi.org/10.31 09/07388551.2015.1132406

Yang, L., Giannis, A., Chang, V. W. C., Liu, B., Zhang, J., & Wang, J. Y. (2015). Application of hydroponic systems for the treatment of source-separated human urine. *Ecological Engineering*, *81*, 182–191. https://doi.org/10.1016/j.ecoleng.2015.04.013

6 Water Quality Monitoring in Wastewater Phytoremediation

6.1 INTRODUCTION

The search for long-term approach to curb and manage water pollution is crucial to the public, policy makers and the government. Water pollution occurs due to persistent inflow of pollutants from nonpoint and point sources into aquatic ecosystems. In general, physical (colour, turbidity, temperature and taste), chemical (phosphate, ammoniacal nitrogen and nitrites) and biological analyses are often used to assess water quality. Furthermore, the biological, chemical and physical properties of water determine its quality and application for a range of purposes, as well as preservation and protection of aquatic habitats (Omer, 2019). The biodegradation and dilution of organic and inorganic compounds allow aquatic environment to adapt and make up for variations in water quality parameters (Loucks & van Beek, 2005).

Nevertheless, hydroponic and built wetland systems for phytoremediation techniques have demonstrated to be an efficient, low-cost and sustainable technology for integrating wastewater treatment and biomass production. Furthermore, hydroponic systems require small space and facilitate the production of valuable crops and plants of high economic importance during wastewater remediation processes. The main advantage of constructed wetlands is that it covers a wide range of applications as it can be used in habitat formation, flood control and wastewater decontamination. Thus, this chapter investigates the efficiency of different plants in tertiary treatment of domestic wastewater using hydroponic systems at different retention times based on different physicochemical analysis.

6.2 CASE STUDY: WATER QUALITY MONITORING IN PHYTOREMEDIATION OF DOMESTIC WASTEWATER

The pH, colour, turbidity, chemical oxygen demand (COD), biochemical oxygen demand (BOD_5), phosphate, ammonia nitrogen and nitrate level of the influent and effluent water samples were determined using the procedures described below. The influent (untreated) samples were collected after secondary treatment, while the effluent (treated) samples were collected from the hydroponic tanks after aquatic plant cultivation.

DOI: 10.1201/9781003359586-6

6.2.1 DETERMINATION OF pH

The test for pH was performed using a pH meter (SCHEMLZ Rev V2.0, TPS Pty Ltd., Australia). The pH probe was stirred in the water samples until a consistent and precise pH value was achieved. Additionally, the pH measurements were conducted at a temperature of 25°C (Instruments, 2000).

6.2.2 DETERMINATION OF COLOUR

The colour test for all the samples was performed using DR3900 spectrophotometer (HACH, Loveland, Co., USA), programmed on 120 colour 455 nm (APHA, 2017). The samples were prepared by filling 10 mL of each of the samples into separate sample cells. The blank sample was used to calibrate the spectrophotometer to zero before inserting the sample cells containing each of the water samples. The readings displayed on the spectrophotometer were recorded. This procedure was repeated thrice for each sample and the average result was recorded.

6.2.3 DETERMINATION OF TURBIDITY

The turbidity tests were conducted using HANA HI 93703 microprocessor turbidimeter that peaked at 890 nm with a range of 0–1000 NTU. The test was performed by filling 25 mL of the samples into separate sample cells. The sample cells containing each of the samples were placed in the turbidimeter and the readings displayed were recorded. This procedure was repeated thrice for each sample and the average outcome was recorded.

6.2.4 DETERMINATION OF BOD_5

The DO difference in the water samples before and after 5 days of incubation is used to determine the BOD in water (Aniyikaiye et al., 2019). DO meter with model HQ40d (HACH, Loveland, Co., USA) portable analyser as described by the dilution method (method 8043) was used to measure the DO of the samples. The samples were prepared by adding 15 mL of the sample BOD bottle and the aerated solution of BOD nutrient was added up to the mark of 300 mL. Thereafter, the probe of the DO meter was inserted in the sample and stirred, the meter was set to measure the $DO_{initial}$ and the procedure was repeated thrice in order to get a constant reading before recording. Furthermore, the samples were incubated at 20°C for 5 days (IS: 3025 P44) and the DO measurement was repeated after the incubation period. The result obtained was recorded as DO_{final}. The BOD_5 concentration was obtained using Equation 6.1 (APHA, 2005):

$$BOD_5 (mg/L) = \frac{DO_{initial} - DO_{final}}{Volume\ of\ sample} \qquad (6.1)$$

where $DO_{initial}$ and DO_{final} are the initial and final dissolved oxygen, respectively.

6.2.5 DETERMINATION OF COD

The COD tests were conducted in accordance to the United States Environmental Protection Agency (USEPA) Reactor Digestion Method (8000) at 610 nm with DR 3900 spectrophotometer (HACH, Loveland, Co., USA) programmed on 430 COD LR detectable range of 3–150 mg/L (HACH, 2014). The COD samples were prepared by transferring 2 mL of the sample into COD digestion reagent. The mixture was then placed in a preheated DRB200 reactor at a temperature of 150°C for 2 hours. After the heating, the mixture was allowed to cool before inserting in the spectrophotometer programmed at 435 COD HR. The results displayed on the spectrophotometer were recorded. The same procedure was used to carry out the COD analysis for all the samples.

6.2.6 DETERMINATION OF PHOSPHATE

The phosphate analysis was performed following USEPA method 365.2 and 4500-P-E standard method for wastewater. The ascorbic acid method (HACH method 8084) was performed using phosVer 3 Phosphate Reagent Powder Pillows with a HACH DR3900 spectrophotometer at 830 nm that allows high-range detection (0.02–2.50 mg/L PO_4^{3-}) programmed on a 490 P React PP (HACH, 2019). The sample was prepared by filling 10 mL of the sample into a sample cell, then one sachet of phosVer 3 Phosphate Reagent Powder was transferred into the sample cell, followed by vigorous shaking of the mixture for 20–30 seconds. After that, the mixture was allowed to settle for 2 minutes before placing it in the sample holder of the spectrophotometer. The results obtained were recorded. The same procedure was used in analysing the phosphate concentration of all the samples.

6.2.7 DETERMINATION OF AMMONIA NITROGEN

The ammonia test of the influent and effluent water samples was analysed following USEPA 350.2 wastewater analysis using the Nessler method (HACH 8038) with a HACH DR3900 spectrophotometer (program 380N, Ammonia Ness) at a wavelength of 425 nm and a noticeable range of 0.02–2.50 mg/L NH_3-N (HACH, 2019). The sample was prepared by filling the mixing cylinder with 25 mL of the sample, three drops of mineral stabilizer and polyvinyl alcohol dispersing agent was transferred into the 25 mL sample and then the mixture was inverted several times, followed by the addition of 1 mL of Nessler reagent. The mixture was inverted severally to obtain a homogenous mixture. Thereafter, the mixture was allowed to settle for 1 minute before transferring 10 mL of the mixture into a sample cell. The 10 mL sample cell was placed in a sample holder in the spectrophotometer. The results displayed were recorded. The same procedure was used in analysing the remaining samples.

6.2.8 DETERMINATION OF NITRATE

The nitrate analysis of the influent and effluent samples was conducted according to Cd reduction method (HACH method 8039) using NitraVer 5 Nitrate Reagent

Pillows with a HACH DR3900 spectrophotometer using program 355 N and Nitrate HR PP at a measurement wavelength of 500 nm and a range of 0.3–30.0 mg/L NO$_3$-N. The samples were prepared by filling 10 mL of the sample into a sample cell, followed by transferring one sachet of NitraVer 5 Nitrate Pillow powder into the sample cell and vigorous shaking of the mixture for 1 minute to obtain a homogenous solution. Thereafter, the sample was allowed to settle for 5 minutes before placing the sample cell into the sample holder of the spectrophotometer. The results obtained were recorded. The same procedure was used in analysing the nitrate concentration of all the samples.

6.2.9 STATISTICAL ANALYSIS

The readings of the physicochemical analysis were recorded in triplicate and the average results were expressed as mean ± standard deviation. Additionally, Microsoft® Excel statistical package was used in calculating the mean and standard deviation. Analysis of variance (ANOVA) IBM® SPSS version 25 package was used in evaluating the significance of difference and t-test (SPSS, 2017). Additionally, Equations 6.2 and 6.3 were used in calculating the dilution factor and percentage reduction efficiency (Valipour et al., 2015).

$$\text{Dilution factor} = \frac{\text{Total volume of diluted sample}}{\text{Volume of wastewater sample}} \tag{6.2}$$

$$\text{Reduction efficiency}\,(\%) = \frac{\text{Ci} - \text{Ce}}{\text{Ci}} \times 100 \tag{6.3}$$

where Ci is the influent concentration and Ce is the effluent concentration.

6.3 OUTCOME OF THE WATER ASSESSMENT OF THE INFLUENT AND EFFLUENT WATER SAMPLES

This section presents and discusses the outcome of the physicochemical analysis performed during the phytoremediation of the domestic wastewater.

6.3.1 ANALYSIS OF COLOUR

Colour assessment is used in determining the quality of wastewater. The concentration of the suspended solids is proportional to the colour of the water sample. The findings of the colour tests performed on the influent and effluent samples are presented in Figures 6.1–6.3.

From Figure 6.1, the colour levels of the influent samples were observed to be within 222 ± 1 to 376 ± 2.6 Pt-Co. Similarly, 376 ± 2.6 Pt-Co colour concentration of the influent sample was reduced to 233.3 ± 1.5 Pt-Co (*P. stratiotes*), 300.6 ± 3.5 Pt-Co (*S. molesta*) and 253.3 ± 3.1 (*E. crassipes*) on the second day of the sampling study. Additionally, up to 50.5% (*P. stratiotes*), 40.5% (*S. molesta*) and 45.4%

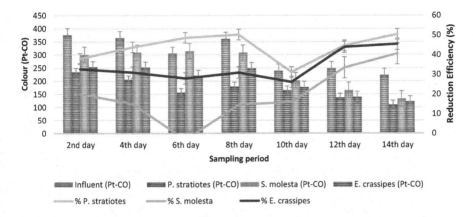

FIGURE 6.1 Graph of colour analysis against sampling days at 6 hours retention time.

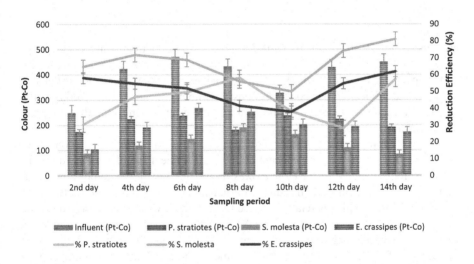

FIGURE 6.2 Graph of colour analysis against sampling days at 12 hours retention time.

(*E. crassipes*) reduction was observed. Furthermore, it was found that *P. stratiotes* performed better than *S. molesta* and *E. crassipes* in lowering the colour content of the influent water at 6 hours retention. Additionally, significant difference ($p < 0.05$) was observed in the ANOVA analysis between the influent and the treatment systems.

According to Figure 6.2, the colour level of the influent samples fluctuated all through the sampling study and was within 249 ± 4.6 to 471.3 ± 1.5 Pt-Co. The treatment systems decreased the colour of the influent samples from 249 ± 4.6 Pt-Co to 173.6 ± 3.2 Pt-Co (*P. stratiotes*), 87.6 ± 3.8 Pt-Co (*S. molesta*) and 104 ± 2.1 Pt-Co (*E. crassipes*) on the second day of the sampling study. Similarly, on the final day, 451 ± 1 Pt-Co colour value of the influent sample decreased to 193 ± 2 Pt-Co (*P. stratiotes*), 85 ± 1 Pt-Co (*S. molesta*) and 172 ± 1 Pt-Co (*E. crassipes*). The maximum reduction

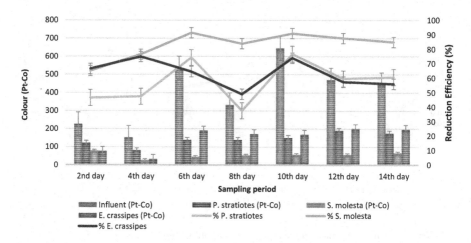

FIGURE 6.3   Graph of colour analysis against sampling days at 24 hours retention time.

efficiencies for *P. stratiotes*, *S. molesta* and *E. crassipes* treatment systems were observed to be 58.1%, 81.1% and 61.9%, respectively. Therefore, *S. molesta* treatment system performed better than *P. stratiotes* and *E. crassipes* treatment systems in reducing the influent colour concentrations at 12 hours retention time. Additionally, the ANOVA analysis between the influent and the effluent samples indicated a significant change ($p < 0.05$).

Figure 6.3 represents the results of the colour analysis at 24 hours retention. From Figure 6.3, the average colour concentration of the influent samples varied throughout the sampling study. In the *P. stratiotes* treatment system, the colour concentration was reduced steadily with reduction efficiency of 76.7% observed on the tenth day of cultivation, which is relatively close to *E. crassipes* with 74% reduction at the same exposure period, whereas in the *S. molesta* treatment system, the reduction took place from the beginning to the final day of the sampling study. A reduction percentage of up to 64.6%–91.4% was obtained from the *S. molesta* treatment system. Similarly, *S. molesta* exhibited the highest reduction efficiency of 91.4% on the sixth day. Additionally, ANOVA analysis revealed a significant difference ($p < 0.05$) between the influent and effluent samples.

Furthermore, it was obvious that the selected plants significantly decreased the colour content of the influent samples. At 6 hours retention time, the overall average colour concentration of 169.3 Pt-Co (*P. stratiotes*), 247.4 Pt-Co (*S. molesta*) and 201.4 Pt-Co (*E. crassipes*) was obtained against 302.2 Pt-Co of the influent samples. For 12 hours retention time, the overall average colour concentration of 210.4 Pt-Co (*P. stratiotes*), 129.2 Pt-Co (*S. molesta*) and 198.0 Pt-Co (*E. crassipes*) was observed as against 397.9 Pt-Co of the influent samples. Similarly, the overall average colour concentration at 24 hours retention time was 141.7 Pt-Co (*P. stratiotes*), 55.7 Pt-Co (*S. molesta*) and 147.6 Pt-Co (*E. crassipes*) against 401.1 Pt-Co of the influent samples. These outcomes indicated that *S. molesta* treatment systems at 12 and 24 hours retention gave the optimum conditions for reducing the colour content of the influent samples. Furthermore, it was found that the initial grey black appearance of the influent samples changed to colourless during the cultivation period. This colour change could be ascribed to the ability of the plant roots to settle, floc, filter and entrap the solid suspended particle in the treatment systems. Additionally, the effluent samples from the *S. molesta* treatment systems meet class IIA water quality standards for Malaysia (DOE, 2005).

6.3.2 ANALYSIS OF TURBIDITY

The results of the turbidity tests performed on the influent and effluent samples are presented in Figures 6.4–6.6.

According to Figure 6.4, it was found that the cultivation of the selected macrophytes decreased the turbidity concentration of the influent samples. The selected plants decreased the turbidity level of the influent sample from 30.66 ± 2.12 to 10.3 ± 0.1 NTU, 15.63 ± 0.06 NTU and 12.36 ± 0.05 NTU for *P. stratiotes*, *S. molesta* and *E. crassipes* treatment systems, respectively. The reduction efficiency ranged from 48% to 72.7% (*P. stratiotes*), 10.19% to 58.3% (*S. molesta*) and 28.3% to 58.88% (*E. crassipes*). Similarly, these outcomes demonstrated that *P. stratiotes* performed better than the other two plants at retention time of 6 hours. Additionally, the average performance exhibited by *S. molesta* could be attributed to the shrinkage and slow growth of the leaves observed from the sixth day onwards. Furthermore, a significant difference ($p < 0.05$) was observed between the influent samples and the individual effluent samples.

From Figure 6.5, the test plants were effective in polishing the influent samples at 12 hours detention time. The turbidity level of the influent samples ranged from 20.7 to 41.8 NTU. However, the cultivation of the selected plants led to the improvement of the influent samples from 28.6 ± 0.45 to 10.7 ± 0.2 NTU (*P. stratiotes*), 4.74 ± 0.12 NTU (*S. molesta*) and 9.02 ± 0.69 NTU (*E. crassipes*) on the fourth day

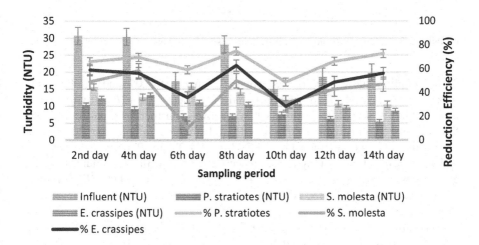

FIGURE 6.4 Graph of turbidity analysis against sampling days at 6 hours retention time.

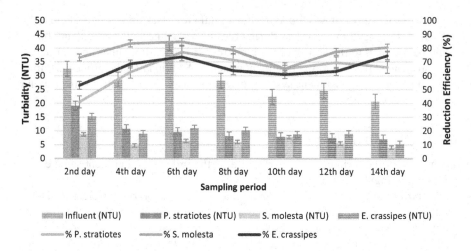

FIGURE 6.5   Graph of turbidity analysis against sampling days at 12 hours retention time.

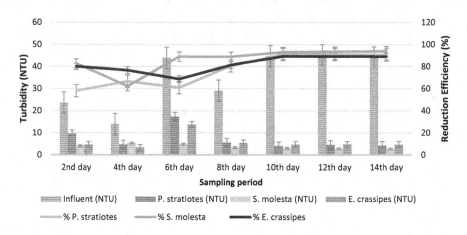

FIGURE 6.6   Graph of turbidity analysis against sampling days at 24 hours retention time.

of the sampling study. Similarly, reduction efficiencies of 41.0% and 77.1% (*P. stratiotes*), 73.2% and 84.6% (*S. molesta*), and 53.0% and 73.6% (*E. crassipes*) were observed on the second and sixth days, respectively. Furthermore, on the 14th day, the treatment systems reduced the turbidity of the influent sample from 20.72 ± 0.06 to 7.0 ± 0.1 NTU, 4.03 ± 0.06 NTU and 5.27 ± 0.27 NTU for *P. stratiotes*, *S. molesta* and *E. crassipes* treatment systems, respectively. Finally, the maximum reduction efficiencies of 66.2% (*P. stratiotes*), 84.6% (*S. molesta*) and 74.6% (*E. crassipes*) were recorded. Additionally, ANOVA tests between the influent and the effluent samples were found to be significant ($p < 0.05$).

According to Figure 6.6, it was revealed that the turbidity concentration of the influent samples was lowered on the second day from 23.63 ± 0.18 to 9.5 ± 1.32 NTU, 4.06 ± 1.45 NTU and 4.72 ± 0.03 NTU for *P. stratiotes*, *S. molesta* and *E. crassipes* treatment systems, respectively. As a result, *P. stratiotes*, *S. molesta* and *E. crassipes* treatment systems recorded reduction efficiencies of 58%, 82.8% and 80%, respectively. The three treatment systems also showed a consistent pattern of decreasing the turbidity as the number of sampling days increases. Furthermore, slight changes were observed on the fourth and sixth days, followed by an onward increase in the reduction efficiency from the *E. crassipes* treatment system. Furthermore, *S. molesta* exhibited the highest removal efficiency of 94%, followed by *P. stratiotes* with 91%, and then *E. crassipes* with 89.3%. Additionally, the ANOVA test showed a significant difference ($p < 0.05$) between the influent and effluent samples.

Turbidity is the cloudiness of water (APHA, 2005), as a result of suspended materials found in water (Alley, 2007). This study assessed the potentials of *P. stratiotes*, *S. molesta* and *E. crassipes* plants in turbidity reduction of the influent water samples at varied detention times of 6, 12 and 24 hours for 2 weeks. At 6 hours retention, the overall average turbidity value was 7.58 NTU (*P. stratiotes*), 12.8 NTU (*S. molesta*) and 9.6 NTU (*E. crassipes*) against 22.8 NTU (influent turbidity), while the average turbidity value of 9.97 NTU (*P. stratiotes*), 6.2 NTU (*S. molesta*) and 9.8 NTU (*E. crassipes*) was observed at 12 hours retention against 28.9 NTU of the influent samples. At 24 hours retention time, the turbidity value of 7.1 NTU (*P. stratiotes*), 3.7 NTU (*S. molesta*) and 5.93 NTU (*E. crassipes*) was recorded against 34.9 NTU of the influent samples. In the overall experiment, up to 93.9% turbidity reduction efficiency was observed. These results demonstrated the effectiveness of *P. stratiotes*, *S. molesta* and *E. crassipes* plants in improving the water quality characteristics of the influent samples. Furthermore, *S. molesta* treatment system at 24 hours retention time provided the ideal conditions for lowering the turbidity level. However, it was discovered that the ability of the *S. molesta* plants in absorbing the pollutants declined at 6 and 12 hours retention times. Besides, the roots were observed to retain and filter the suspended solids via rhizofiltration. The long hairy roots of the selected test plants facilitated the rhizofiltration process. Root hairs have an electrical current which aids in digestion and assimilation of the suspended particles by the plants (Johnson, 1993). Additionally, phytoremediation process using shallow hydroponic system improved the clarity of the influent samples by providing a conducive environment for the trapped suspended particle to sediment at the ground level of the pond. The polished/filter water remained at the top level of the pond. These results demonstrated the efficacy of the test plants in refining the influent samples to permitted Malaysian water standard of class I (5 NTU) from the 10th to the 14th days at 24 hours retention time. Similarly, Aswathy (2017) reported up to 85.66% turbidity reduction efficiency in phytoremediation of kitchen wastewater using *P. stratiotes* after 10 days detention time. Additionally, Parwin and Paul (2019) recorded up to 86.75% turbidity removal in wastewater remediation using *E. crassipes*.

6.3.3 ANALYSIS OF pH

pH value of water samples indicates the level of the acidity or alkalinity. It is used to determine the relative quantity of hydroxyl ions and free hydrogen in water. The results obtained from the pH tests of the influent and effluent samples at retention times of 6, 12 and 24 hours are presented in Figures 6.7–6.9.

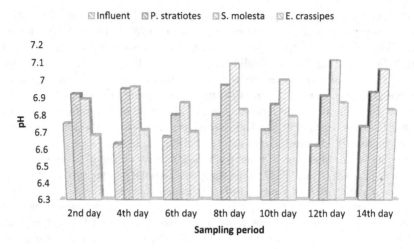

FIGURE 6.7 Graph of pH against sampling days at 6 hours retention time.

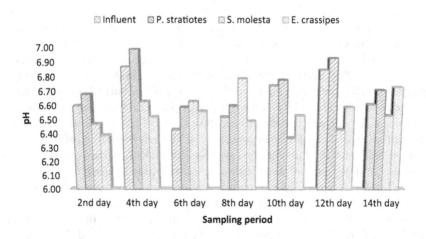

FIGURE 6.8 Graph of pH against sampling days at 12 hours retention time.

According to Figure 6.7, the pH values of the influent samples were observed to be within 6.62 ± 0.01 to 6.80 ± 0.03. It was found that the cultivation of the aquatic plants in the influent samples adjusted the pH level of the effluent samples. Additionally, the ANOVA analysis revealed a statistically significant difference ($p < 0.05$) between the influent samples and the *P. stratiotes* and *S. molesta* treatment systems, whereas no significant difference ($p > 0.05$) was found between the influent samples and *E. crassipes* treatment system at 6 hours retention time.

From Figure 6.8, the pH analysis of the influent samples was observed to be in the range of 6.43 ± 0.02 to 6.87 ± 0.01. Additionally, little change was recorded in the pH values of the effluent water samples. Furthermore, the pH of the effluent samples was

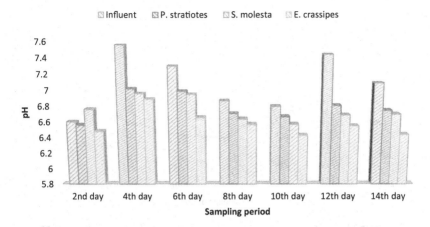

FIGURE 6.9 Graph of pH against sampling day at 24 hours retention time.

found to be neutral (6.59–6.99) during the sampling period. Similarly, the ANOVA test gave a statistically significant difference ($p < 0.05$) between the influent samples and the *P. stratiotes* treatment systems, while the ANOVA test between the influent water samples and *S. molesta* and *E. crassipes* treatment systems revealed no significant change ($p > 0.05$).

According to Figure 6.9, the pH of the influent samples was found to be within 6.6 ± 0.05 to 7.56 ± 0.01 throughout the 14 days sampling study. In addition, the selected plants demonstrated a similar pattern in the pH value of the influent samples, fluctuating between the values of 6.43 ± 0.02 and 7.02 ± 0.05. On the fourth day, *P. stratiotes* treatment system recorded a pH of 7.02 ± 0.05 as against pH 7.56 ± 0.40 of the influent sample. The ANOVA analysis between the pH of the influent and the effluent samples from the three treatment systems revealed a significant difference ($p < 0.05$).

Additionally, the overall outcome showed a consistent pH in the range of 6.30–7.11. The macrophytes improved the pH, thereby enhancing the plant roots and microorganisms to successfully break down the residual contaminants in the influent samples. Thus, this research corroborates with the findings of Rezania et al. (2015), who reported an optimum pH value of 6.6–8.0. Furthermore, the treatment system at 24 hours retention time is the optimal pH condition for phytoremediation of the domestic wastewater. Similarly, the pH values observed from this research is within the class I (6.5–8.5) of Malaysian water quality standards (DOE, 2005). Besides, the outcome observed in this study is in agreement with the research of Mahmood et al. (2005) and Priya and Selvan (2017). Furthermore, Haidara et al. (2018) used 100 g of *E. crassipes* and *P. stratiotes* in remediation of aquaculture wastewater at 21 days retention time. The outcome of their study gave a pH of 6.37 ± 0.09 to 6.53 ± 0.24.

6.3.4 ANALYSIS OF COD

Chemical oxygen demand (COD) can be described as the quantity of oxygen required to decompose organic matter in water. It uses chemical characteristics to estimate the

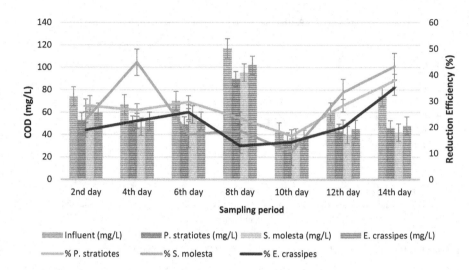

FIGURE 6.10   Graph of COD analysis against sampling days at 6 hours retention time.

level of pollution in water (Tchobanoglous et al., 2003). The outcomes of the COD tests performed on the influent and effluent water samples are presented in Figures 6.10–6.12.

From Figure 6.10, 74 ± 0 mg/L of the influent COD value was lowered to 53 ± 0, 67 ± 0 and 60 ± 0 mg/L by *P. stratiotes*, *S. molesta* and *E. crassipes*, respectively, on the second day of the sampling study. In addition, reduction efficiencies of 28.3%, 23% and 18.9% were obtained by *P. stratiotes*, *S. molesta* and *E. crassipes*, respectively. Furthermore, up to 37.8% (*P. stratiotes*), 44.8% (*S. molesta*) and 35.1% (*E. crassipes*) COD removal was obtained on the 14th day at 6 hours retention time. Additionally, the ANOVA analysis between the influent and effluent samples indicated a significant difference ($p < 0.05$), but no significant difference ($p > 0.05$) was observed between the influent and *S. molesta* effluent samples.

According to Figure 6.11, the COD of the influent samples was inconsistent throughout the sampling duration at 12 hours retention time. At the beginning of the sampling period, *P. stratiotes*, *S. molesta* and *E. crassipes* actively lowered the COD of the influent sample from 55.3 ± 0.57 to 46.3 ± 0.5, 30 ± 0 and 38 ± 0 mg/L, correspondingly. This change indicated COD reduction of 16.27% (*P. stratiotes*), 45.75% (*S. molesta*) and 31.2% (*E. crassipes*). On the 12th day, a noteworthy decrease was detected as the COD of the influent samples was reduced from 90 ± 0 to 46 ± 0 mg/L (*P. stratiotes*), 25 ± 1 mg/L (*S. molesta*) and 35 ± 1 mg/L (*E. crassipes*). Furthermore, up to 74.8%, 72.2% and 61.1% COD reduction was obtained from *P. stratiotes*, *S. molesta* and *E. crassipe*s treatment systems, respectively. The ANOVA analysis revealed a significant difference ($p < 0.05$) between the influent and the effluent samples.

According to Figure 6.12, it was obvious that the cultivation of the test plants decreased the COD level of the influent samples. On the second day, COD reduction of 5.88% (*P. stratiotes*), 64.7% (*S. molesta*) and 66.9% (*E. crassipes*) was observed.

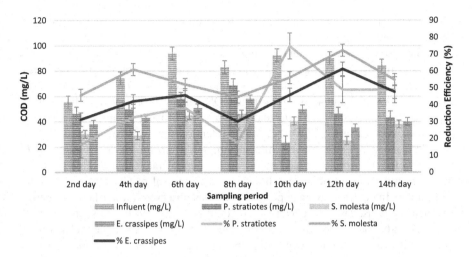

FIGURE 6.11 Graph of COD analysis against sampling days at 12 hours retention time.

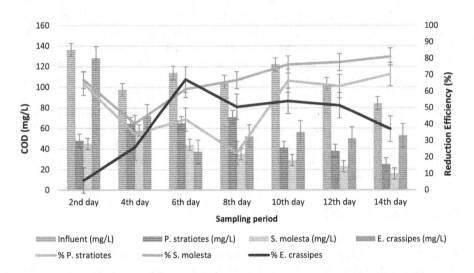

FIGURE 6.12 Graph of COD analysis against sampling days at 24 hours retention time.

After that, fluctuations were observed on the fourth, sixth and eighth days in which a lower reduction COD percentage was obtained for *P. stratiotes* effluents. Similarly, the maximum COD reduction efficiency for *P. stratiotes* was obtained on the last day with a value of 70.3%. In the case of *S. molesta*, the COD reduction trend maintained a stable increase from the 6th to 14th day of the sampling study. The maximum COD reduction efficiency for *S. molesta* was observed on the last day of the sampling study, with a value of 81%, whereas the trend for *E. crassipes* effluent samples continued to fluctuate between 50.6% and 67.2% from the 6th day to the 12th day.

Furthermore, a significant difference ($p < 0.05$) was observed between the influent and the effluent samples.

Furthermore, at 6 hours retention time, 71.28, 53.65, 58.41 and 58.71 mg/L was obtained as the overall average COD value of the influent samples and *P. stratiotes*, *S. molesta* and *E. crassipes* effluent samples, respectively. For 12 hours retention time, 81.8, 49.9, 36.2 and 45 mg/L average COD value was observed for the influent samples, *P. stratiotes*, *S. molesta* and *E. crassipes* effluent samples, respectively, while the overall average COD at 24 hours retention time was recorded as 108.84, 50.3, 35.7 and 64.0 mg/L for the influent samples, *P. stratiotes*, *S. molesta* and *E. crassipes* effluent samples. Thus, these results demonstrated that *S. molesta* treatment systems provided better conditions for the removal of COD from the influent samples. In contrast, average performance was observed in the 6 hours retention treatment systems. The low performance could be attributed to the short retention time and high load of the pollutants contained in the influent samples, which overwhelmed the roots of the test plants, thereby making them weak to absorb the excess nutrients. In comparison, concerning retention times, the 24 hours treatment systems show a better performance than the other two treatment systems as up to 80% COD reduction was achieved from the *S. molesta* plants. Additionally, the reduction in COD might be due to the oxidation of organic carbon to carbon dioxide and water. The possibility of COD breakdown after a few days by microbes is highly likely in the roots of the plant (Ng et al., 2017). The metabolic activity of aerobic microorganisms in wastewater may be heightened by utilizing the accessible organic matter as a substrate (Wickramasinghe & Jayawardana, 2018). This may result in the breakdown of the organic loads in the influent samples and also a subsequent decrease in COD. Additionally, oxygen supplied through the plant roots helps to degrade organic matter (Sooknah, 2000). Furthermore, the COD level of the effluent samples falls within the class III (25–50) water quality standards for Malaysia (DOE, 2005). Additionally, Lu et al. (2018) reported COD removal of 61.70% (*P. stratiotes*) and 68.2% (*E. crassipes*) in the phytoremediation of polluted water.

6.3.5 ANALYSIS OF BOD$_5$

The amount of biodegradable compounds in water and wastewater can be determined through BOD$_5$ analysis. It estimates the quantity of oxygen required for aerobic microorganisms to break down organic matter in water (Carr & Neary, 2008). The results of BOD$_5$ analysis performed on the influent and effluent water samples are presented in Figures 6.13–6.15.

From Figure 6.13, it was found that the BOD$_5$ of the influent samples varied throughout the sampling study. On the second day, the treatment systems reduced the BOD$_5$ of the influent sample from 14.6 to 11.6 mg/L (*P. stratiotes*), 13.6 mg/L (*S. molesta*) and 11.9 mg/L (*E. crassipes*). This decrease led to reduction efficiencies of 20.5%, 6.8% and 17.9% for *P. stratiotes*, *S. molesta* and *E. crassipes* treatment systems, respectively. Furthermore, a slow increase in the reduction efficiency of BOD$_5$ was obtained in the *P. stratiotes* effluent samples from the 20.5% recorded on the second day to 38.9%, 40%, 36.6% and 46.7% on the fourth, sixth, eighth and tenth days of the sampling period, respectively, while for the *S. molesta* effluent samples,

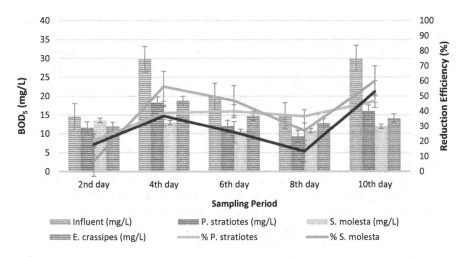

FIGURE 6.13 Graph of BOD_5 analysis against sampling period at 6 hours retention time.

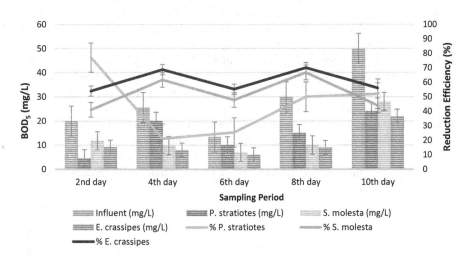

FIGURE 6.14 Graph of BOD_5 analysis against sampling days at 12 hours retention time.

the BOD_5 removal efficiency increased from the 6.8% observed on the second day to 56.4%, 47%, 27.1% and 60% on the fourth, sixth, eighth and tenth days, respectively. The removal efficiency by *E. crassipes* increased from the 17.9% observed on the second day to 36.9%, 26%, 13.4% and 53% on the fourth, sixth, eighth and tenth days, respectively. Additionally, among the three treatment systems, *S. molesta* performed better than *P. stratiotes* and *E. crassipes* in lowering the BOD_5 concentration of the influent samples. Furthermore, a significant difference ($p < 0.05$) was observed between the influent and the individual effluent samples.

From Figure 6.14, it is obvious that the BOD_5 value of the influent samples was reduced from 20 to 4.6 mg/L (*P. stratiotes*), 11.8 mg/L (*S. molesta*) and 9.2 mg/L

(*E. crassipes*) on the second day of the sampling study at 12 hours retention time. Furthermore, a slow decrease of the BOD_5 level was obtained in the *P. stratiotes* treatment system on the fourth and sixth days with 21.1% and 25.4% reduction efficiency, respectively, while for the *S. molesta* effluent samples, the removal efficiency increased from the 41.1% observed on the second day to 61.7%, 47.8%, 66.7% and 44% on the fourth, sixth, eighth and tenth days, respectively. The removal efficiency by *E. crassipes* increased from the 54% observed on the second day to 68.8%, 55.2%, 70% and 56% on the fourth, sixth, eighth and tenth days, respectively. Furthermore, a significant change ($p < 0.05$) was observed between the influent and the individual effluent samples.

According to Figure 6.15, it was evident from the graph that the BOD_5 value of the influent samples was reduced from 38.7 to 33 mg/L and 29 mg/L by *P. stratiotes* and *S. molesta*, respectively, on the second day. Similarly, up to 15.7% (*P. stratiotes*), 24.5% (*S. molesta*) and 4.6% (*E. crassipes*) BOD_5 reduction was recorded on the fourth day. On the eighth day, the treatment system reduced the BOD_5 level of the influent samples from 37.2 to 17.4 mg/L (*P. stratiotes*), 9.4 mg/L (*S. molesta*) and 15.6 mg/L (*E. crassipes*). Additionally, *S. molesta* recorded the highest reduction efficiency of 74.73%, whereas *E. crassipes* and *P. stratiotes* treatment systems recorded up to 58% and 53.2%, respectively. A significant difference ($p < 0.05$) was observed between the influent and effluent treatment systems.

Furthermore, the findings obtained from the BOD_5 analysis indicated that the selected plants were capable of reducing the BOD_5 level of the influent samples. Additionally, the cultivation of the selected plants in the domestic water sample increased the DO level, which led to lowering of the BOD_5 concentration. In addition, the pH of 6–8 obtained during the cultivation period enhanced the BOD_5 reduction in the influent samples. Similarly, optimum condition for lowering the BOD_5 of the influent samples was observed in *S. molesta* treatment system at 12 hours retention time. In general, *S. molesta* demonstrated better tendency for reduction of the

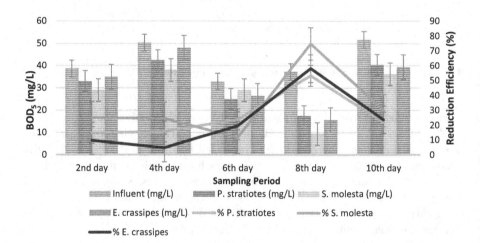

FIGURE 6.15 Graph of BOD_5 analysis against sampling days at 24 hours retention time.

BOD concentration than *P. stratiotes* and *E. crassipes*. The increase in DO of water enhanced the photosynthetic activity, thereby increasing aerobic bacterial activity that led to degradation of COD and BOD in contaminated water (Saha et al., 2017). Moreover, DO is necessary for aquatic species and it enters the water by diffusion from the atmosphere, and as a by-product of the photosynthesis process by algae and aquatic plants. Furthermore, DO concentration of less than 2 mg/L indicates poor water quality and may have a detrimental effect on aquatic life (Gorde & Jadhav, 2013; Hazmi & Hanafiah, 2018).

6.3.6 ANALYSIS OF PHOSPHATE

Phosphates are primarily found in domestic wastewater as a result of detergents, food waste, anthropogenic and animal activities (Kroiss et al., 2011). Orthophosphate (PO_4) in wastewater is a necessary nutrient for plant growth. Macrophytes absorb phosphates via root assimilation into shoots and xylem, thereby decreasing the phosphate content in wastewater (Lambers & Colmer, 2005). The findings of the phosphate tests performed on the influent and effluent samples are presented in Figures 6.16–6.18.

From Figure 6.16, it was found that the phosphate value of the influent samples ranged from 4.21 to 6.8 mg/L. The phosphate level in the influent samples steadily decreased all through the sampling period. Similarly, the phosphate level of the influent samples decreased from 6.8 ± 0 to 5.40 ± 0.02 mg/L (*P. stratiotes*), 5.55 ± 0.01 mg/L (*S. molesta*) and 5.44 ± 0 mg/L (*E. crassipes*) on the second day. This led to reduction efficiency of 20.6% (*P. stratiotes*), 18.7% (*S. molesta*) and 20% (*E. crassipes*). On the sixth day, the influent phosphate values were lowered from 4.26 ± 0 to 3.38 ± 0 mg/L

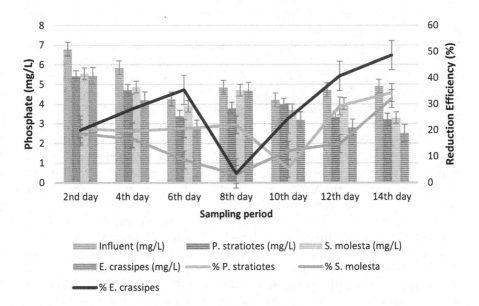

FIGURE 6.16   Graph of phosphate analysis against sampling days at 6 hours retention time.

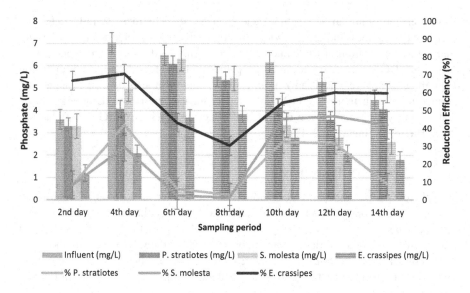

FIGURE 6.17   Graph of phosphate analysis against sampling days at 12 hours retention time.

(*P. stratiotes*), 3.89±0 mg/L (*S. molesta*) and 2.75±0 mg/L (*E. crassipes*). Additionally, a maximum phosphate removal efficiency of 34.3% (*P. stratiotes*), 32% (*S. molesta*) and 48.7% (*E. crassipes*) was achieved on the 14th day. Furthermore, the ANOVA analysis between the influent and the treatment systems indicated a significance difference ($p < 0.05$). Finally, the test plants performed slowly in lowering the phosphate concentration of the influent samples, with a removal efficiency of less than 50% attained during the 14-day cultivation period.

From Figure 6.17, it was evident the selected plants lowered the phosphate level of the influent samples from 3.6±0.04 to 3.3±0.04 mg/L (*P. stratiotes*), 3.3±0 mg/L (*S. molesta*) and 1.2±0.04 mg/L (*E. crassipes*), while on the eighth day, the test plants reduced phosphate concentration of the influent samples from 5.56±0.05 to 5.36±0.04 mg/L (*P. stratiotes*), 5.44±0 mg/L (*S. molesta*) and 3.84±0 mg/L (*E. crassipes*). Besides, the maximum phosphate reduction of 70.5% was recorded for *E. crassipes* on the second day, followed by *S. molesta* (47%) on the 12th day and *P. stratiotes* (32.4%) on the tenth day. Additionally, slow reduction performance of below average was observed in *S. molesta* and *P. stratiotes* treated wastewater samples. Furthermore, the ANOVA analysis of the influent and effluent samples indicated a significant difference ($p < 0.05$).

From Figure 6.18, the phosphate levels in the influent samples were found to be variable, ranging from 4 to 20.1 mg/L. A steady phosphate reduction was observed throughout the sampling period. Similarly, the selected plants decreased the 20.01±0.01 mg/L phosphate level of the influent sample to 10.50±0.01 mg/L (*P. stratiotes*), 13.16±0.01 mg/L (*S. molesta*) and 6.53±0.15 mg/L (*E. crassipes*), while on the eighth day, the selected plants decreased the phosphate concentration from 5.56±0.05 to 2.76±0.05 mg/L (*P. stratiotes*), 0.96±0.01 mg/L (*S. molesta*)

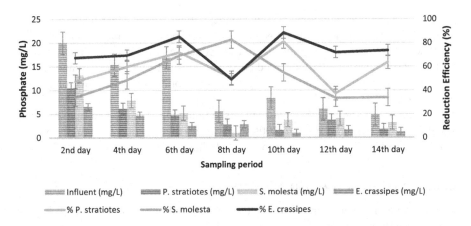

FIGURE 6.18 Graph of phosphate analysis against sampling days at 24 hours retention time.

and 2.83 ± 0.05 mg/L (*E. crassipes*). The highest phosphate reduction efficiency for *P. stratiotes* and *E. crassipes* treatment systems was observed on the tenth day with a value of 81.2% and 88.5%, respectively, while up to 82.7% phosphate reduction was observed for *S. molesta* on the eighth day. Furthermore, the ANOVA test demonstrated a significance change ($p < 0.05$) between the influent and the treatment systems except for *S. molesta*, which was found to be not significant ($p > 0.05$).

In addition, at 6 hours retention time, the overall average phosphate reduction of 3.97 mg/L (*P. stratiotes*), 4.29 mg/L (*S. molesta*) and 3.65 mg/L (*E. crassipes*) was obtained against the average influent phosphate concentration of 5.08 mg/L, while at 12 hours retention time, the overall average phosphate concentration of 4.38 mg/L (*P. stratiotes*), 4.11 mg/L (*S. molesta*) and 2.5 mg/L (*E. crassipes*) was recorded against the average influent phosphate level of 5.50 mg/L. Similarly, the overall average phosphate values of 4.46 mg/L (*P. stratiotes*), 5.45 mg/L (*S. molesta*) and 2.92 mg/L (*E. crassipes*) were observed at 24 hours retention time against the average influent phosphate level of 11.01 mg/L. These outcomes demonstrated that the selected plants reduced the phosphate content in the influent samples at varying degrees. The natural differences in macronutrient needed for physiological mechanisms may have led to the alterations demonstrated by the three plants (Tisdale et al., 1993). Further, *E. crassipes* was more effective compared to *P. stratiotes* and *S. molesta* plants in absorbing the excess phosphate present in the influent samples. Furthermore, the results of this current research showed that the 6 and 12 hours treatment system did not provide favourable conditions for the plant roots to effectively absorb and assimilate the phosphate compounds. Hence, it is clear from these findings that retention time is a crucial factor that influences and regulates the rate of nutrient uptake and assimilation by plants in phytoremediation systems.

Furthermore, plants require phosphorus as a macronutrient. It is a major component of adenosine triphosphate (ATP) and adenosine diphosphate (ADP), both of which are required for energy storage. In addition, it helps in the transport of energy during photosynthesis and production of ADP and ATP, as well as the synthesis of their biochemical components (nucleotides, sugar and nucleic acids) (Tisdale et al., 1993). Hence, free orthophosphate was absorbed from the influent samples over the 14-day cultivation period. Similarly, reduction of nitrogen and phosphorous by *E. crassipes* was similarly testified by Wickramasinghe and Jayawardana (2018). Moreover, the findings of this investigation corroborate with the work of Nizam et al. (2020), who reported that water hyacinth performed better than *P. stratiotes*, *I. aquatica*, *C. asiatica* and *S. molesta* plants in reduction of phosphate from polluted water. Henry-Silva and Camargo (2006) reported up to 82% (*E. crassipes*), 83.3% (*P. stratiotes*) and 72.1% (*S. molesta*) reduction of phosphorous.

6.3.7 ANALYSIS OF AMMONIA NITROGEN

The outcome of the ammonia nitrogen tests performed on the water samples is presented in Figures 6.19–6.21.

From Figure 6.19, the ammonia nitrogen of the influent samples was observed to be inconsistent and ranged from 12.8 to 23.89 mg/L. During the 14-day treatment period, a gradual decrease in the influent ammonia nitrogen level was observed. Additionally, the ammonia nitrogen removal efficiency was observed to be 12.9%–20.4% (*P. stratiotes*), 4.2%–14.1% (*S. molesta*) and 4%–35.9% (*E. crassipes*) at 6 hours retention time. Furthermore, the ANOVA test revealed a statistically significant change ($p < 0.05$) between the influent and effluent samples.

From Figure 6.20, the ammoniacal nitrogen level of the influent samples falls within 13.4–21.86 mg/L. Similarly, on the second day, the test plants reduced the nitrogen concentration of the influent samples from 13.4 ± 0.1 to 12.5 ± 0.1 mg/L (*P. stratiotes*), 13.04 ± 0.01 mg/L (*S. molesta*) and 10.1 ± 0 mg/L (*E. crassipes*). The changes showed a little reduction efficiency of 6.71% for *P. stratiotes*, 2.68% for *S. molesta* and 24.6% for *E. crassipes*. Furthermore, the 23.2 ± 0.04 mg/L ammonia nitrogen of the influent sample was drastically reduced to 8.0 ± 0 mg/L (*P. stratiotes*),

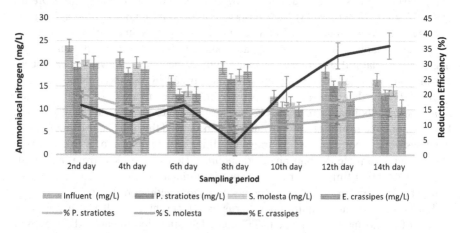

FIGURE 6.19 Graph of ammoniacal nitrogen analysis against sampling days at 6 hours retention time.

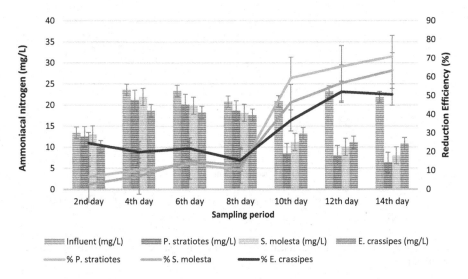

FIGURE 6.20 Graph of ammoniacal nitrogen analysis against sampling days at 12 hours retention time.

10.04 ± 0.01 mg/L (*S. molesta*), and 11.13 ± 0.05 mg/L (*E. crassipes*) on the 12th day, while on the 14th day, the ammoniacal nitrogen value of the influent sample was lowered from 21.86 ± 0 to 6.36 ± 0.05 mg/L (*P. stratiotes*), 8.01 ± 0.02 mg/L (*S. molesta*) and 10.8 ± 0.1 mg/L (*E. crassipes*). Additionally, the ammoniacal nitrogen removal efficiency at 12 hours retention time for *P. stratiotes*, *S. molesta* and *E. crassipes* was estimated to be in the range of 60%–71%, 45%–65% and 36%–50%, respectively. Similarly, *P. stratiotes* performed better than *S. molesta* and *E. crassipes*. Meanwhile, the ANOVA test revealed a statistically significant change ($p < 0.05$) between the influent samples and the individual effluent samples.

According to Figure 6.21, it is obvious that the effluent samples indicated a constant rate of reduction in the ammoniacal nitrogen from the beginning to the last day of the sampling study with a little inconsistency on the fourth day. Similarly, it was observed that the treatment systems reduced the ammoniacal nitrogen of the influent samples from 19.9 ± 0.17 to 11.04 ± 0.25 mg/L (*P. stratiotes*), 14.8 ± 0.1 mg/L (*S. molesta*) and 8.2 ± 0.1 mg/L (*E. crassipes*) on the second day. On the 14th day, the ammoniacal nitrogen of the influent samples was lowered from 22.05 ± 0.01 to 2.5 ± 0.08 mg/L (*P. stratiotes*), 2.1 ± 0 mg/L (*S. molesta*) and 2.4 ± 0.17 mg/L (*E. crassipes*). Furthermore, the ANOVA analysis revealed a statistically significant difference ($p < 0.05$) between the influent and the individual effluent samples.

Furthermore, the overall results showed that about 80%–90% of ammonia nitrogen removal was achieved. Similarly, the overall average ammonia nitrogen reduction of 15.10 mg/L (*P. stratiotes*), 16.33 mg/L (*S. molesta*) and 14.76 mg/L (*E. crassipes*) was achieved against 18.23 mg/L average ammonia nitrogen of the influent samples at 6 hours retention time, while for the 12 hours retention time, the overall average ammonia nitrogen reduction of 13.62 mg/L (*P. stratiotes*), 14.60 mg/L (*S. molesta*)

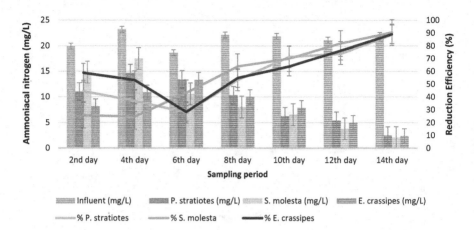

FIGURE 6.21 Graph of ammoniacal nitrogen analysis against sampling days at 24 hours retention time.

and 14.26 mg/L (*E. crassipes*) was recorded against 20.01 mg/L average ammonia nitrogen of the influent samples. Furthermore, an overall average ammonia nitrogen concentration of 9.08 mg/L (*P. stratiotes*), 9.05 mg/L (*S. molesta*) and 8.25 mg/L (*E. crassipes*) was observed at 24 hours retention time against 21.26 mg/L average ammonia nitrogen level of the influent samples. Additionally, the highest ammonia nitrogen reduction efficiency on the 14th day of the sampling study was recorded as 88.7% (*P. stratiotes*), 90.5% (*S. molesta*) and 89.1% (*E. crassipes*) at 24 hours retention time. Thus, the lowering efficiency trend for the three treatment systems followed the pattern of 24 hours > 12 hours > 6 hours. Furthermore, the results of this investigation suggest that the 6 hours retention time was insufficient for the test plants to absorb or break down the ammonia nitrogen contained in the influent samples. Additionally, plants take up nutrients through direct and indirect (microbial activity at rhizosphere) ways, and they play important roles in absorbing ammoniacal nitrogen from wastewater (Deng & Ni, 2013; Wenwei et al., 2016). Microbial activity helps in elimination of nitrogen via plant absorption, since it accelerates nitrogen breakdown at the rhizosphere, increases the activity of other elements and improves bioavailability (Ting et al., 2018). Additionally, the findings obtained in the 12 and 24 hours retention time support the findings published by Nivetha et al. (2016), who found that *P. stratiotes* reduced up to 84.8% ammonia nitrogen from sewage wastewater. Furthermore, Parwin and Paul (2019) discovered that *E. crassipes* removed up to 94.36% of ammonia nitrogen from wastewater.

6.3.8 ANALYSIS OF NITRATE

The findings of the nitrate tests performed on the influent and effluent samples are presented in Figures 6.22–6.24.

From Figure 6.22, the nitrate level of the influent samples falls within 4.36–8.7 mg/L. The uptake rate of nitrate by the treatment systems increased across the

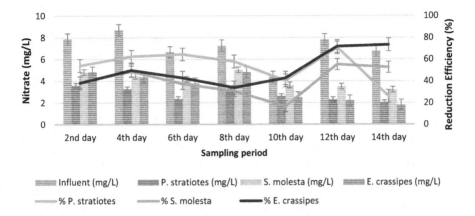

FIGURE 6.22 Graph of nitrate analysis against sampling days at 6 hours retention time.

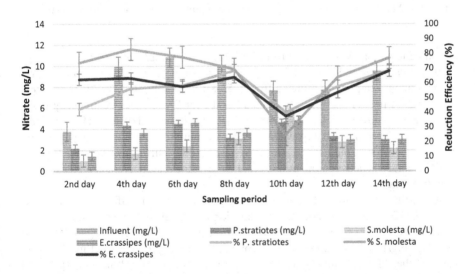

FIGURE 6.23 Graph of nitrate analysis against sampling days at 12 hours retention time.

14 days cultivation period. On the second day, 7.86 ± 0.05 mg/L nitrate value of the influent was reduced to 3.6 ± 0 mg/L (*P. stratiotes*), 4.8 ± 0 mg/L (*S. molesta*) and 4.83 ± 0 mg/L (*E. crassipes*). Similarly, 7.81 ± 0.02 mg/L nitrate level of the influent sample was reduced to 2.3 ± 0.1 mg/L for *P. stratiotes*, 3.5 ± 0.1 mg/L for *S. molesta* and 2.2 ± 0 for *E. crassipes* effluent samples. Additionally, the maximum reduction efficiency of 77.6% was recorded by *P. stratiotes* on the 12th day, whereas the maximum reduction efficiencies of 55.3% and 73.3% were obtained for *S. molesta* and *E. crassipes*, respectively, on the 14th day. Additionally, the ANOVA analysis revealed a significant difference ($p < 0.05$) between the influent and effluent samples.

According to Figure 6.23, the nitrate levels of the influent samples ranged within 3.8–10.83 mg/L. On the second day of the experiment, the nitrate level of

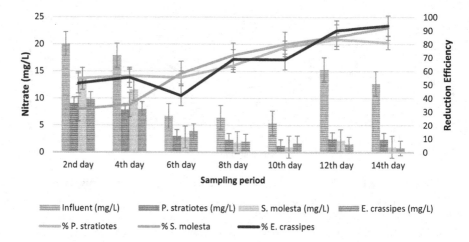

FIGURE 6.24 Graph of nitrate analysis against sampling days at 24 hours retention time.

the influent sample was reduced from 3.8 ± 0 to 2.2 ± 0 mg/L (*P. stratiotes*), 1.0 ± 0 mg/L (*S. molesta*) and 1.43 ± 0.05 mg/L (*E. crassipes*). Similarly, the nitrate level of the influent sample was reduced from 9.52 ± 0.10 to 3.01 ± 0.02 mg/L for *P. stratiotes*, 2.2 ± 1 mg/L for *S. molesta* and 3.04 ± 0.01 for *E. crassipes* treatment systems. The highest removal efficiencies of 68.38% (*P. stratiotes*) and 68.06% (*E. crassipes*) were recorded on the 14th day, whereas the maximum reduction efficiency of 82.9% was recorded for *S. molesta* treatment system on the fourth day. Additionally, the ANOVA analysis between the influent and the treatment systems indicated a significant difference ($p < 0.05$).

According to Figure 6.24, the average nitrate level of the influent samples was within 5.4–20 mg/L. The nitrate concentration of the influent samples was lowered from 20 ± 0.1 to 9.05 ± 0.01 mg/L (*P. stratiotes*), 13.6 ± 0 mg/L (*S. molesta*) and 9.8 ± 0.17 mg/L (*E. crassipes*) on the second day. Similarly, the 5.4 ± 0.1 mg/L nitrate value of the influent sample was decreased to 1.2 ± 0.1 mg/L (*P. stratiotes*), 1.0 ± 0 mg/L (*S. molesta*) and 1.7 ± 0.1 mg/L (*E. crassipes*). Furthermore, the lowest and highest nitrate removal efficiencies for *P. stratiotes* were recorded on the 2nd and 12th days with a value of 54.75% and 83.6%, respectively, while the lowest and highest nitrate removal efficiencies for *S. molesta* were observed to be 32% and 92.1% on the 2nd and 14th days, respectively. In addition, the minimum and maximum percentage reduction efficiency for *E. crassipes* was found to be 41% and 93% on the 6th and 14th days of the experiment, respectively. Moreover, the ANOVA analysis between the influent and the treatment systems revealed a significant difference ($p < 0.05$).

The overall outcome demonstrated that the test plants were effective in lowering the nitrate content in the influent samples. Similarly, the overall average nitrate reduction of 2.74 mg/L (*P. stratiotes*), 4.1 mg/L (*S. molesta*) and 3.47 mg/L (*E. crassipes*) against 7.05 mg/L average nitrate of the influent samples was obtained at 6 hours retention time, while the overall average nitrate reduction of 3.6 mg/L (*P. stratiotes*),

2.71 mg/L (*S. molesta*) and 3.45 mg/L (*E. crassipes*) was recorded at 12 hours retention time against 8.52 mg/L average nitrate reduction of the influent samples. The overall average nitrate values of 4.03 mg/L (*P. stratiotes*), 4.85 mg/L (*S. molesta*) and 3.95 mg/L (*E. crassipes*) were obtained for the 24 hours retention time against 12.05 mg/L average nitrate reduction of the influent samples. Furthermore, the 24 hours retention time demonstrated a better nitrate uptake performance than the 6 and 12 hours detention time. Additionally, the findings of this investigation revealed that *S. molesta* and *E. crassipes* plants exhibited the highest efficiency in nitrate uptake compared to *P. stratiotes* plants at 12 and 24 hours retention time. Similarly, the fast growth observed in the three test plants could be due to high nitrate assimilations at 12 and 24 hours retention times. At 12 hours, *S. molesta* (82.9%) performed better than *E. crassipes* (68.38%) and *P. stratiotes* (68.06%), while the trend of the nitrate reduction efficiency for the three test plants at 6 hours retention time was in the order of *P. stratiotes* > *E. crassipes* > *S. molesta*. Furthermore, the influent samples were polished to class I from the 6th day to the 14th day at 24 hours retention time. In other words, macrophytes have long fibrous roots that offer a large surface area for microbial, chemical and physical activities for nitrification and nutrient absorption (Akinbile et al., 2015). Additionally, the findings obtained at 24 hours retention time coincide with the nitrate reduction results reported by Nivetha et al. (2016). Ingersoll and Baker (1998) recorded nitrate removal of up to 90% for *E. crassipes*, 51% for *P. stratiotes* and 36% for *S. molesta* in phytoremediation of synthetic wastewater. Ayyasamy et al. (2009) reported up to 61%–83% nitrate removal by *E. crassipes* from contaminated water, although the findings from this study are not in support of the research by Ng et al. (2017), in which phytoremediation of wastewater using *S. molesta* yielded up to 36% (phosphate), 19% (nitrate) and 31% (ammonia).

6.4 CONCLUSION

This chapter presented the outcome of research on the potential of selected macrophytes in phytoremediation of wastewater. The results obtained from this research prove that the three aquatic plants were efficient in tertiary treatment of wastewater at different retention times. Thus, the lowering efficiency trend for the three treatment systems followed the pattern of 24 hours > 12 hours > 6 hours. Furthermore, the results of this investigation suggested that the 6 hours retention time was insufficient for the test plants to absorb or break down the ammonia nitrogen contained in the influent samples. In most cases, the highest removal efficiencies were recorded on the 12th and 14th days. Therefore, it is clear from these findings that retention time is a crucial factor that influences and regulates the rate of nutrient uptake and assimilation by plants in phytoremediation systems.

REFERENCES

Akinbile, C. O., Ogunrinde, T. A., Che Bt Man, H., & Aziz, H. A. (2015). Phytoremediation of domestic wastewaters in free water surface constructed wetlands using Azolla pinnata. *International Journal of Phytoremediation*, *18*(1), 54–61. https://doi.org/10.1080/1522 6514.2015.1058330

Alley, E. (2007). *Water Quality Control Handbook*. McGrawHill, New York.

Aniyikaiye, T. E., Oluseyi, T., Odiyo, J. O., & Edokpayi, J. N. (2019). Physico-chemical analysis of wastewater discharge from selected paint industries in Lagos, Nigeria. *International Journal of Environmental Research and Public Health*, *16*(1235), 1–17. https://doi.org/10.3390/ijerph16071235

APHA. (2005). *Standard Methods* (21st ed.). American Public Health Association, Washington.

APHA. (2017). Color, true and apparent. *Standard Methods for the Examination of Water and Wastewater and National Council for Air and Stream Improvement (NCASI) Methods*, *30*(11), 2771–2775.

Aswathy, M. (2017). Wastewater treatment using constructed wetland with water lettuce (Eichornia crasipies). *International Journal of Civil Engineering and Technology*, *8*(8), 1413–1421.

Ayyasamy, P. M., Rajakumar, S., Sathishkumar, M., Swaminathan, K., Shanthi, K., Lakshman-aperumalsamy, P., & Lee, S. (2009). Nitrate removal from synthetic medium and groundwater with aquatic macrophytes. *Desalination*, *242*(1–3), 286–296. https://doi.org/10.1016/j.desal.2008.05.008

Carr, G. M., & Neary, J. P. (2008). *Water Quality for Ecosystem and Human Health* (2nd ed.). United Nations Environment Programme (UNEP), Nairobi, Kenya.

Deng, Y., & Ni, F. (2013). Review of ecological floating bed restoration in polluted water. *Journal of Water Resource and Protection*, *5*(12), 1203–1209. https://doi.org/10.4236/jwarp.2013.512128

Department of Environment (DOE). (2005). *Environmental Quality Report, 2002–2005, Malaysia*. Ministry of Science, Technology and Environment, Malaysia.

Gorde, S. P., & Jadhav, M. V. (2013). Assessment of water quality parameters: A review. *International Journal of Engineering Research and Applications*, *3*(6), 2029–2035.

HACH. (2014). Chemical oxygen demand, Method 8000. *Hach*, DOC316.53., 10. http://www.hach.com/asset-get.download.jsa?id=7639983816

HACH. (2019). *Phosphorous, Reactive (Orthophosphate) Method 8048 (DOC316.53.01119)* (10th ed.). Water Analysis Handbook, Hach Company, Loveland, CO.

Haidara, A. M., Magami, I. M., & Sanda, A. (2018). Bioremediation of aquacultural effluents using hydrophytes. *Bioprocess Engineering*, *2*(4), 33–37. https://doi.org/10.11648/j.be.20180204.11

Hazmi, N. I. A., & Hanafiah, M. M. (2018). Phytoremediation of livestock wastewater using *Azolla filiculoides and Lemna minor*. *Environment & Ecosystem Science*, *2*(1), 13–16. https://doi.org/10.26480/ees.01.2018.13.16

Henry-Silva, G. G., & Camargo, A. F. M. (2006). Efficiency of aquatic macrophytes to treat Nile tilapia pond effluents. *Scientia Agricola*, *63*(5), 433–438. https://doi.org/10.1590/s0103-90162006000500003

Ingersoll, T. L., & Baker, L. A. (1998). Nitrate removal in wetland microcosms. *Water Research*, *32*(3), 677–684. https://doi.org/10.1016/S0043-1354(97)00254-6

Instruments, H. (2000). *Hi 2210 hi 2211 Microprocessor-based pH/mV/oC Bench Meters*. www.hannainst.com

Johnson, C. (1993). Mechanism of Water Wetland Water Quality Interaction. In G. A. Moshiri (Ed.), *Constructed Wetland for Water Quality Improvement* (pp. 7). Lewis Publishers, Boca Raton, FL.

Kroiss, H., Rechberger, H., & Egle, L. (2011). Phosphorus in Water Quality and Waste Management. In Sunil Kumar (Eds), *Integrated Waste Management* (Vol. 2, pp. 181–214). Intechopen, London, United Kingdom.

Lambers, H., & Colmer, T. D. (2005). Root physiology – from gene to function. *Plant Soil*, *274*, 7–15.

Loucks, D. P., & van Beek, E. (2005). Water Quality Modeling and Prediction. In *Water Resource Systems Planning and Management*. Springer, Cham. https://doi.org/10.1007/978-3-319-44234-1_10

Lu, B., Xu, Z., Li, J., & Chai, X. (2018). Removal of water nutrients by different aquatic plant species: An alternative way to remediate polluted rural rivers. *Ecological Engineering*, *110*, 18–26. https://doi.org/10.1016/j.ecoleng.2017.09.016

Mahmood, Q., Zheng, P., Siddiqi, M. R., Islam, E. U., Azim, M. R., & Hayat, Y. (2005). Anatomical studies on water hyacinth (Eichhornia crassipes (Mart.) Solms) under the influence of textile wastewater. *Journal of Zhejiang University: Science*, *6B*(10), 991–998. https://doi.org/10.1631/jzus.2005.B0991

Ng, Y. S., Samsudin, N. I. S., & Chan, D. J. C. (2017). Phytoremediation capabilities of *Spirodela polyrhiza* and *Salvinia molesta* in fish farm wastewater: A preliminary study. *IOP Conference Series:* Materials Science *and Engineering*, *206*(1). https://doi.org/10.1088/1757-899X/206/1/012084

Nivetha, C., Subraja, S., Sowmya, R., & Induja, N. M. (2016). Water lettuce for removal of nitrogen and phosphate from sewage. *IOSR Journal of Mechanical and Civil Engineering (IOSR-JMCE)*, *13*(2), 104–107. https://doi.org/10.9790/1684-13020198101

Nizam, N. U. M., Hanafiah, M. M., Noor, I. M., & Karim, H. I. A. (2020). Efficiency of five selected aquatic plants in phytoremediation of aquaculture wastewater. *Applied Sciences (Switzerland)*, *10*(8), 1–11. https://doi.org/10.3390/APP10082712

Omer, N. A. (2019). Water Quality Parameters. In J. Kevin Summers (Ed.), *Water Quality - Science, Assessments and Policy* (p. 19). IntechOpen, London, United Kingdom. https://doi.org/https://doi.org/10.5772/intechopen.89657

Parwin, R., & Paul, K. K. (2019). Phytoremediation of kitchen wastewater using Eichhornia crassipes. *Journal of Environmental Engineering*, *145*(6), 1–10. https://doi.org/10.1061/(ASCE)EE.1943-7870.0001520

Priya, S., & Selvan, S. (2017). Water hyacinth (Eichhornia crassipes) – An efficient and economic adsorbent for textile effluent treatment – A review. *Arabian Journal of Chemistry*, *10*, S3548–S3558. https://doi.org/10.1016/j.arabjc.2014.03.002

Rezania, S., Din, M., Taib, S., Dahalan, F., Songip, A., Singh, L., & Hesam, K. (2015). The efficient role of aquatic plant (water hyacinth) in treating domestic wastewater in continuous system. *International Journal of Phytoremediation*. https://doi.org/10.1080/15226514.2015.1130018

Saha, P., Shinde, O., & Sarkar, S. (2017). Phytoremediation of industrial mines wastewater using water hyacinth. *International Journal of Phytoremediation*, *19*(1), 87–96. https://doi.org/10.1080/15226514.2016.1216078

Sooknah, R. (2000). A review of the mechanisms of pollutant removal in water hyacinth systems. *Science and Technology - Research Journal*, *6*, 49–57.

SPSS. (2017). *ANOVA SPSS Package* (No. 25). IBM® SPSS version 25 package.

Tchobanoglous, G., Burton, F., Stensel, H., & Metcalf, E. (2003). Wastewater Engineering: Treatment and Reuse (4th ed.). Tata McGraw-Hill, New York.

Ting, W. H. T., Tan, I. A. W., Salleh, S. F., & Wahab, N. A. (2018). Application of water hyacinth (Eichhornia crassipes) for phytoremediation of ammoniacal nitrogen: A review. *Journal of Water Process Engineering*, *22*(October), 239–249. https://doi.org/10.1016/j.jwpe.2018.02.011

Tisdale, S. L., Nelson, W., Beaton, J. D., & Havlin, J. (1993). *Elements Required in Plant Nutrition: Soil Fertility and Fertilizers*. MacMillan Publishing Co, New York.

Valipour, A., Raman, V. K., & Ahn, Y. H. (2015). Effectiveness of domestic wastewater treatment using a bio-hedge water hyacinth wetland system. *Water (Switzerland)*, *7*(1), 329–347. https://doi.org/10.3390/w7010329

Wenwei, W., Ang, L., Konghuan, W., Lei, Z., Xiaohua, B., Kun-zhi, L., Ashraf, M. A., & Limei, C. (2016). The physiological and biochemical mechanism of nitrate-nitrogen removal by water hyacinth from agriculture eutrophic wastewater. *Brazilian Archives of Biology and Technology*, *59*(Special Issue 1), 1–10. https://doi.org/10.1590/1678–4324–2016160517

Wickramasinghe, S., & Jayawardana, C. K. (2018). Potential of aquatic macrophytes Eichhornia crassipes, Pistia stratiotes and Salvinia molesta in phytoremediation of textile wastewater. *Journal of Water Security*, *4*, 1–8. https://doi.org/10.15544/jws.2018.001

7 Water Quality Monitoring Using Internet of Things (IoT)

7.1 INTRODUCTION

About 2 billion people worldwide use the internet to browse, exchange information, social networking and many other applications. The internet is a breakthrough for connecting with machines and intelligent objects to collect, interpret, compute and coordinate data (Miorandi et al., 2012). IoT is a fast-emerging soft computing technology that is used in industries for monitoring and management processes. The advancement of smart infrastructure and the proliferation of IoT provides opportunities in wastewater sector for the collection and reporting of real-time data on sewage conditions for operation management (Edmondson et al., 2018). However, due to the complexities and sensitivity of the wastewater treatment processes, there is a need to apply different approaches to obtain accurate and real-time conditions of wastewater quality. Real-time data access can be obtained using remote IoT monitoring technology such as Arduino wireless sensor devices. Data collected at monitoring sites can be visualized on a server computer (Chowdury et al., 2019). In the aspect of water quality monitoring, wireless sensor devices provide significant benefits in terms of communication, installation time and reduced operational cost (Priyadharshini et al., 2018).

7.2 INTERNET OF THINGS (IoT) IN WASTEWATER MONITORING

IoT has been used as an integrated system in applications ranging from smart power grids to smart wearables, intelligent supply chains and smart cities (Thombre et al., 2016). IoT has attracted tremendous attention in establishing a network for sending and receiving information through sensors. This invention has the capacity to enhance the operation of large sewage systems. IoT can be used in environmental research, water quality monitoring and controlling system, monitoring of air pollution and snow level, detection of early earthquake and forest fire, and prevention of avalanche and landslides (Andersson & Hossain, 2014). Additionally, IoT environmental sensor is a valuable tool for monitoring and modelling environmental phenomena (Edmondson et al., 2018). Additionally, sensors are connected to a network of different chipboards such as the Arduino microprocessors, ESP32, ESP8266 and ARM-based Raspberry pi microcomputer to collect data that can be visualized and evaluated using a computer and artificial intelligence (AI) models (Chowdury et al., 2019). For example, Chen and Han (2018) monitored the conductivity, dissolved oxygen (DO), salinity, temperature, presumptive Enterococci, *E. coli*, total coliforms and faecal streptococci in river water samples using a sensor device. The results obtained indicated that the integrated system was capable of collecting and displaying the water quality information online. The success of the study proved that IoT devices could detect probable sources of pollutants in water. Carminati et al. (2020) designed a low-cost sensor technology used for early warning, predictive maintenance and efficient control of water assessment parameters. Additionally, IoT data has a temporal component and may be modelled as a time series. Additionally, data from water quality monitoring is highly complicated, dynamic and non-stationary.

7.3 HARDWARE DESIGN OF THE ARDUINO IoT SYSTEM

Arduino is an open source software and hardware application that excels in interpreting and analysing data from a variety of sources to carry out programmed operations. The coding language of the Arduino system is easy to implement in various projects. In addition, the Arduino system is inexpensive, capable of managing multiple tasks, simplifies complicated tasks and is compatible with all desktop operating systems. Furthermore, the software and hardware of the Arduino systems are constantly expanded and enhanced with code libraries and third-party boards. The Arduino code is separated into three sections; variables are defined and third-party libraries are incorporated at the start of the code. The set-up, which is in the second section, commands the information sources used in the code to start. The last section is called the loop, which controls and runs the code. Furthermore, the Arduino board consists of sensor nodes made up of PIC microcontroller, liquid crystal display (LCD) and sim card, which are used in determining the water level, turbidity, total dissolved solid (TDS), temperature, oxidation reduction potential (ORP), pH and gases. The Arduino Board UNO is connected to the internet through an external Wi-Fi module while the sensor nodes are placed in the water samples. The information gathered by the sensors is shown on the LCD and transmitted straight to the Open Source IoT Website on computer (Priyadharshini et al., 2018). Geetha and Gouthami (2017) developed a smart water quality monitoring system coupled with Wi-Fi module controller to measure the conductivity, pH and turbidity of water samples. Furthermore, Qin et al. (2018) designed an electrochemically modified water monitoring device consisting of sensing probes with inkjet-printed pH, temperature and chlorine sensors. Figure 7.1 illustrates the connection diagram of IoT system.

7.4 SENSOR NODES OF THE ARDUINO IoT SYSTEM

7.4.1 TEMPERATURE SENSOR

DF Robot temperature sensor is mostly used in measuring the temperature of the surrounding. Temperature sensors are classified into different categories, including thermometers, register temperature detectors, thermocouples, infrared semiconductors, solid-state temperature sensors and thermistors. The sensing capability of these sensors varies from one sensor to the other (Paul, 2018). A thermistor is made up of resistor that varies with temperature. Due to its resistive nature, the voltage between the terminals must be read using an excitation source. In most cases, thermistor

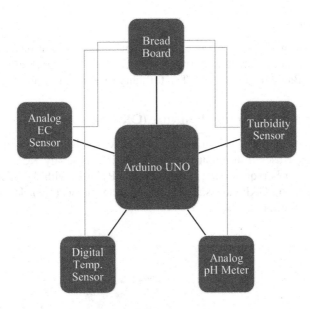

FIGURE 7.1   Connection diagram of IoT system.

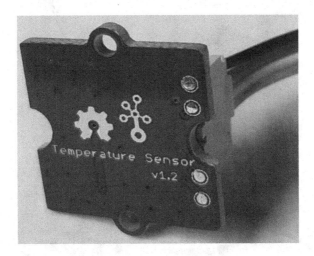

FIGURE 7.2   DF Robot temperature sensor.

temperature sensor from DF Robot is used because it has greater design control and can be utilized for applications below 300°C, making it suitable for systems that works at ambient temperatures. The voltage is proportional to the temperature through a positive temperature coefficient (PTC) or negative temperature coefficient (NTC). Thermistor temperature sensors have been used for different applications due to their portability and accuracy (Cloete et al., 2016). The DF Robot temperature sensor is shown in Figure 7.2.

7.4.2 Turbidity Sensor

Turbidity sensor generates voltage output through measurement of light scattering caused by suspended particles. Turbidity is measured in Nephelometric Turbidity Units (NTU) (Baldwin et al., 2019). The turbidity sensor is shown in Figure 7.3.

7.4.3 Oxidation Reduction Potential (ORP) Sensor

ORP sensors are similar to pH sensor and in some cases they are combined together. The ORP sensor uses a different reference solution, typically KCL Ag/Ag-Cl. A signal conditioner is required to interface the ORP sensor with the microcontroller (Cloete et al., 2016). ORP values are expressed in millivolts (mV). The ORP sensor is shown in Figure 7.4.

FIGURE 7.3 Turbidity sensor.

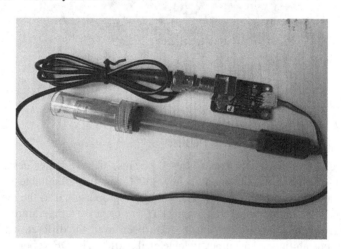

FIGURE 7.4 Oxidation reduction potential (ORP) sensor.

7.4.4 TOTAL DISSOLVED SOLIDS (TDS) SENSOR

TDS sensor is used to measure the dissolved solids in water samples. TDS sensor can be connected to IoT systems for easy application (Anuradha et al., 2018). Additionally, TDS sensors accept input and output voltage ranging from 3.3–5.5V and 0–2.3V, respectively, making them suitable with 3.3 or 5 V control systems. The excitation source is an alternating current signal that prevents the probe from being polarized and extends the durability of the probe. TDS has a measuring range of 0–1000 ppm (Baldwin et al., 2019). TDS sensor is waterproof and can be used in hydroponic systems and wastewater treatment plants (WWTP) monitoring (Anuradha et al., 2018). The TDS sensor is shown in Figure 7.5.

7.5 LIQUID CRYSTAL DISPLAY (LCD)

LCD device is a flat panel that applies the light-modulating characteristics of liquid crystals and polarizers to display different characters and symbols. LCDs consist of 16-character and 2-line alphanumeric (16×2) LCD interface linked to the microcontroller, which enables the device to connect to E-block I/O ports. The data on the LCD is in serial format and can be powered by a 5 V source. The optimum source of 5 V is the E-block's Multi programmer or 5 V fixed power source (Priyadharshini et al., 2018). The LCD is shown in Figure 7.6.

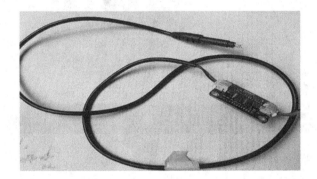

FIGURE 7.5   Total dissolve solids (TDS) sensor.

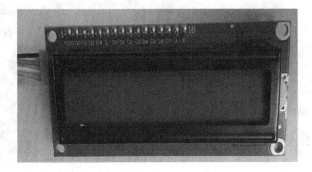

FIGURE 7.6   Liquid crystal display (LCD).

7.6 WI-FI MODULE

Wi-Fi is a wireless network that connects devices to the internet. Wi-Fi compatible devices can be linked to cyberspace through LAN web and wireless point within the range of about 20 m indoors and a larger compass outdoors to allow the internet reach within the capability of an internet-connected wireless meshwork (Chowdury et al., 2019). The Wi-Fi module is shown in Figure 7.7.

7.7 GLOBAL SYSTEM MOBILE (GSM SHIELD)

GSM modules are available in a variety of shapes and sizes, each having a particular function. The IoT board consists of SIM900 GPRS modem for internet connectivity and controller that converts all input UART data to GPRS-based online data. The module is interfaced with Arduino mega and it is controlled by AT command through UART. It requires 12 V power source from an AC 100–240 V to DC 12 V 2A power adapter (Paul, 2018). The GSM compartment is shown in Figure 7.8.

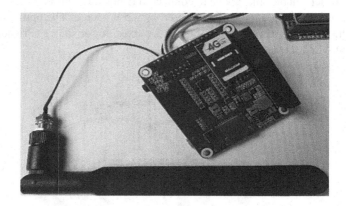

FIGURE 7.7 Wi-Fi module.

FIGURE 7.8 Arduino board showing the GSM compartment.

FIGURE 7.9   IDE and sketch (codes).

7.8   CODING DEVELOPMENT OF THE IoT SYSTEM

The Arduino integrated development environment (IDE) is used in writing and uploading the code into the microcontroller. Arduino code is also called sketches. The sketch is processed and assembled into machine language. The environment is written in C++ and includes a number of custom methods and functions. After the Arduino code compilation, it is uploaded into the board memory. The IDE and sketch (codes) are shown in Figure 7.9.

7.9   CONCLUSION

IoT systems have been used in several industries for control functions and process monitoring. IoT-based water quality monitoring is superior to traditional methods because it is eco-friendly, fast, cheap, sustainable and does not require harmful chemicals. Furthermore, the incorporation of IoT systems in wastewater treatment systems such as phytoremediation techniques would enable efficient monitoring of nutrient uptake and early detection of contamination. Additionally, the application of IoT devices in water quality monitoring would act as a big leap for researchers, government agencies and stakeholders in reporting, regulating and policy making of waste discharge laws that would safeguard natural water bodies.

REFERENCES

Andersson, K., & Hossain, M. S. (2014). Smart risk assessment systems using belief-rule-based DSS and WSN technologies. *4th International Conference on Wireless Communications, Vehicular Technology, Information Theory and Aerospace and Electronic Systems, VITAE 2014-Co-Located with Global Wireless Summit*, 1–5. https://doi.org/10.1109/VITAE.2014.6934397

Anuradha, T., Bhakti, Chaitra, R., & Pooja, D. (2018). IoT based low cost system for monitoring of water quality in real time. *International Research Journal of Engineering and Technology*, 5(5), 1658–1663.

Baldwin, B., Kolli, V., Lehman, K., Li, A., Lin, C., Malhotra, R., & Devaraj, H. (2019). Integrating electronics with solid structures using 3D circuits. *2019 IEEE MIT Undergraduate Research Technology Conference, URTC 2019* (2019 IEEE MIT Undergraduate Research Technology Conference, URTC 2019). Institute of Electrical and Electronics Engineers Inc., 1–11.

Carminati, M., Turolla, A., Mezzera, L., Di Mauro, M., Tizzoni, M., Pani, G., Zanetto, F., Foschi, J., & Antonelli, M. (2020). A self-powered wireless water quality sensing network enabling smart monitoring of biological and chemical stability in supply systems. *Sensors (Switzerland)*, 20(1125), 1–16. https://doi.org/10.3390/s20041125

Chen, Y., & Han, D. (2018). Water quality monitoring in smart city: A pilot project. *Automation in Construction*, 89(February), 307–316. https://doi.org/10.1016/j.autcon.2018.02.008

Chowdury, M. S. U., Emran, T. B., Ghosh, S., Pathak, A., Alam, M. M., Absar, N., Andersson, K., & Hossain, M. S. (2019). IoT based real-time river water quality monitoring system. *Procedia Computer Science*, 155, 161–168. https://doi.org/10.1016/j.procs.2019.08.025

Cloete, N. A., Malekian, R., & Nair, L. (2016). Design of smart sensors for real-time water quality monitoring. *IEEE Access*, 4, 3975–3990. https://doi.org/10.1109/ACCESS.2016.2592958

Edmondson, V., Cerny, M., Lim, M., Gledson, B., Lockley, S., & Woodward, J. (2018). A smart sewer asset information model to enable an 'Internet of Things' for operational wastewater management. *Automation in Construction*, 91, 193–205. https://doi.org/10.1016/j.autcon.2018.03.003

Geetha, S., & Gouthami, S. (2017). Internet of things enabled real time water quality monitoring system. *Smart Water*, 2(1), 1–19. https://doi.org/10.1186/s40713-017-0005-y

Miorandi, D., Sicari, S., De Pellegrini, F., & Chlamtac, I. (2012). Internet of things: Vision, applications and research challenges. *Ad Hoc Networks*, 10(7), 1497–1516. https://doi.org/10.1016/j.adhoc.2012.02.016

Paul, B. (2018). *Sensor based water quality monitoring system*. http://hdl.handle.net/10361/10840

Priyadharshini, N. R., Vanishree, R., & Sebasteenav, P. R. (2018). Smart water quality management system. *Global Research and Development Journal for Engineering | National Conference on Advancement in Emerging Technologies (NCAET'18)*, March, 25–29.

Qin, Y., Alam, A. U., Pan, S., Howlader, M. M. R., Ghosh, R., Hu, N. X., Jin, H., Dong, S., Chen, C. H., & Deen, M. J. (2018). Integrated water quality monitoring system with pH, free chlorine, and temperature sensors. *Sensors and Actuators, B: Chemical*, 255, 781–790. https://doi.org/10.1016/j.snb.2017.07.188

Thombre, S., Ul Islam, R., Andersson, K., & Hossain, M. S. (2016). Performance analysis of an IP based protocol stack for WSNs. *IEEE Conference on Computer Communications Workshops (INFOCOM WKSHPS)*, 360–365. https://doi.org/10.1109/INFCOMW.2016.7562102

8 Machine Learning Techniques in Water Quality Monitoring

8.1 INTRODUCTION

Environmental issues and inadequate resources have promoted interest in searching for cost-effective strategies for water resource management. Data-driven and physical methods have been used in forecasting of water quality (WQ) parameters, estimation of sediment load and discharge in hydrological researches. The physical approach, on the other hand, has certain limitations, such as high cost and long time requirements. Currently, machine learning (ML) techniques have been effectively deployed in water treatment plants to complement analytical methods in monitoring, measurement and forecasting of WQ index. Additionally, ML tools are capable of modelling and forecasting complicated non-linear systems and historical data within a short period. Thus, this chapter describes the concept of ML tools in WQ monitoring. Additionally, this chapter describes the development, applications, advantages and disadvantages of machine learning techniques such as artificial neural network (ANN), support vector machine (SVM), adaptive neuro-fuzzy inference system (ANFIS), error ensemble learning approach and classical method (multilinear regression (MLR)) in the evaluation and prediction of water quality models. Furthermore, case studies on applications of ANN, SVM, ANFIS and MLR in phytoremediation of wastewater were explored.

8.2 CONCEPT OF MACHINE LEARNING (ML) TECHNIQUES IN PHYTOREMEDIATION OF WASTEWATER

Artificial intelligence (AI) is a subfield in computer science that deals with the simulation of human intellect in computers (Bagheri et al., 2019). Machine learning (ML) is a branch of AI and computer science that applies data and algorithms to perform classification, pattern recognition and prediction on data by learning from existing data. ML techniques have been used by different researchers because of their simplicity, accuracy and high operational speed (Rajaee et al., 2019). Furthermore, AI approaches have been employed in different engineering fields due to their ability to address practical wastewater treatment challenges (Al Aani et al., 2019; Fan et al., 2018), monitor water quality (Elkiran et al., 2019) and water resource management (Xu et al., 2019; Zhao et al., 2020). Additionally, many studies have used AI models in monitoring and modelling water assessment parameters including turbidity, temperature, pH, chemical oxygen demand (COD), suspended solids (SS), biochemical oxygen demand (BOD) and ammonia nitrogen in water treatment plants without taking into account complex reaction mechanism (Hamed et al., 2004). Besides, it is preferable to use multiple AI models rather than a single AI model for modelling of water quality parameters (Nadiri et al., 2018). Similarly, ML has been successfully used in complex computing tasks where programming algorithms with acceptable results are difficult (Wernick et al., 2010). Furthermore, ML is divided into supervised and unsupervised learning. Supervised learning is used to develop a fundamental understanding of the relationship between input and output values, whereas unsupervised learning does not provide known outcomes to the learning algorithms. As a result, they are left to discover their structures in the inputs (Bagheri et al., 2019). ML algorithms such as ANFIS, ANN (Zhu et al., 2018), DLNN (Shi & Xu, 2018), SVM (Guo et al., 2015) and decision tree are the modern black box that are being applied in the area of water management operations (Pang et al., 2019; Turunen et al., 2018).

8.3 ARTIFICIAL NEURAL NETWORK (ANN)

ANN is a mathematical model that is used to process information in a synaptic weight and brain-like learning process (Gaya et al., 2014). Generally, ANN comprises of tightly coupled processing units called neurons that collaborate to analyse data in a manner comparable to the central nervous system (Hayder et al., 2014). Therefore, the development of ANN model involves the collection and pre-processing of data, data handling, network training and testing, data splitting, selection of inputs, selection of activation function and learning algorithm (Abrahart et al., 2012). Furthermore, the activation function could be linear, hyperbolic tangent sigmoid, Gaussian or logistic sigmoid depending on the selected learning algorithm and scaling technique (Bennett et al., 2013; Esfe, 2017). The most frequently used activation functions in water and wastewater quality monitoring are sigmoid functions such as hyperbolic tangent functions and logistic sigmoid, which can be calculated using the mathematical expressions given in Equations 8.1 and 8.2 (Maier & Dandy, 2000; Oyebode & Stretch, 2019). Additionally, ANN can be classified based on learning techniques as Levenberg-Marquardt (LM), back-propagation (BP), adaptive back-propagation, Bayesian regularization (BR) and gradient descent with momentum learning rate (Rajaee et al., 2019). In most circumstances, when compared to other models, ANN models trained using LM learning methods are faster (Senthil Kumar et al., 2013). Additionally, ANN can be applied in numerical paradigms for function approximation owing to their great characteristics of adaptability, improved input and output mapping, fault tolerance, nonlinearity and self-learning (Abiodun et al., 2018).

$$f(x) = \left[\frac{2}{1 + e^{-2(x)}} \right] - 1 \qquad (8.1)$$

$$f(x) = \left[\frac{1}{1 + e^{-(x)}} \right] \qquad (8.2)$$

where x and $f(x)$ refer to the weighted sum of the inputs and the output of the neuron, respectively.

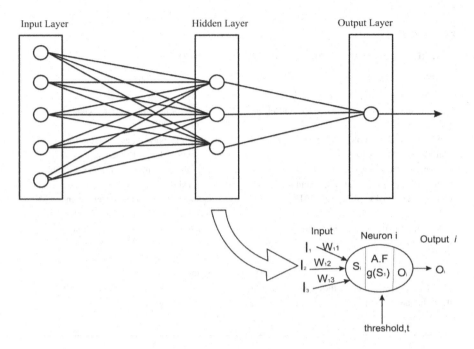

FIGURE 8.1 ANN–FFNN structure showing the input, hidden and output layers. Adapted from Nourani et al. (2018).

ANN has different structures and training algorithms including FFNN, recurrent neural networks and cascade-forward neural networks (CFNN). Additionally, neurons receive input signals, process them, and then transmit the results to all associated artificial neurons (Bagheri et al., 2019). The hidden node processes the data fed into the input layer and transmits the prediction to the output layer during the ANN process. Each training input is transferred to the output layer, where the training error is propagated back to the network until a satisfactory output is obtained (Gayaa et al., 2014). Among the several ANN architectures, the three-layer FFNN is the most frequently employed (Haykin, 1999). Figure 8.1 presents the ANN–FFNN network structure (Adapted from Nourani et al. 2018). Additionally, the use of ANN methods has both advantages and drawbacks. Table 8.1 highlights the advantages and disadvantages of ANN techniques.

8.4 SUPPORT VECTOR MACHINE (SVM)

SVM achieves high generalization capability through the use of structural risk minimization (SRM) induction principle (Vapnik, 1998). SVM is a type of supervised learning algorithm that is used in regression and binary classification. They fall under a class of machine learning techniques known as kernel methods and they are sometimes referred to as kernel machines. SVM was initially designed to solve classification issues, but its use has extended to include regression-type applications for function estimation (Raghavendra & Deka, 2014). SVMs are similar to

TABLE 8.1

Advantages and Disadvantages of ANN Techniques

Advantages	Disadvantages
ANN does not require prior understanding of the structure or connections between the controlling parameters (Sharghi et al., 2018).	ANN is incapable of producing explicit models since their operations are not clear (Basheer & Hajmeer, 2000).
ANN can forecast complex hydrological processes (Adeyemo et al., 2018).	The optimal network architecture for each model may differ (Basheer & Hajmeer, 2000).
In comparison to other modelling tools, ANN methods are in relatively low demand for computational tasks (Dinu et al., 2017).	It is challenging to prioritize an acceptable model since there are no standard criteria to regulate the design and development of ANN model (Basheer & Hajmeer, 2000).
ANN requires little knowledge of the data or scenario to give a satisfactory result (Sharghi et al., 2018).	The inability of ANN to incorporate prior information derived from physical principles serves as a constraint on their use (Basheer & Hajmeer, 2000).
ANN models exhibit compactness, model structure flexibility and the capacity to self-adapt to a given dataset (Sharghi et al., 2018).	They are prone to over-fitting and over-parameterization issues (Adeyemo et al., 2018).

multilayer perceptron (MLP) and radial basis (RBF), since the weight of the network is obtained by solving quadratic programming problem with linear inequality constraints (Dibike et al., 2000).

Despite that SVM has only been used for a short time, it has proven to be a reliable and competent algorithm for regression and classification in different fields. SVMs are used in regression modelling to estimate a single output variable from a collection of input variables (Khan & Coulibaly, 2006). The advantage of this technique is that the prediction accuracy of SVM outperforms many other approaches including decision trees, nearest neighbours and neural networks. Furthermore, SVM algorithms involves the application of the principle of Fermat (1638), principle of Lagrange (1788) and Principle of Kuhn-Tucker (1951) (Raghavendra & Deka, 2014). Additionally, SVM algorithms can better be understood through four fundamental concepts: the soft-margin SVM, the hard-margin SVM, kernel function and separation hyperplane (Raghavendra & Deka, 2014). Figures 8.2 and 8.3 illustrate the SVM algorithms, Adapted from Raghavendra and Deka (2014). Therein, the advantages and disadvantages of SVM are described in Table 8.2.

Furthermore, the fundamental principle of the SVM execution in pattern recognition is the linear or non-linear mapping of the input vectors into a potentially higher dimension of feature space. The type of kernel function determines the mapping process. Then, an optimal hyperplane is built to achieve a maximum separation of the two classes. In other words, SVM training was developed to address the issue of over-fitting and it excels at processing a large number of features (Vapnik, 1998). The network architecture of SVM algorithm is presented in Figure 8.4, Adapted from Nourani et al. (2018).

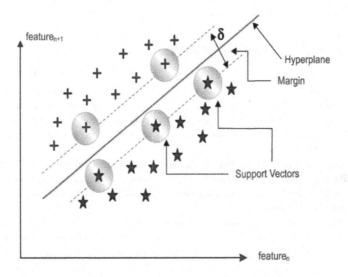

FIGURE 8.2   Maximum separation hyperplane. Adapted from Raghavendra and Deka (2014).

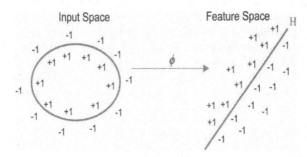

FIGURE 8.3   Linear separation in feature space. Adapted from Raghavendra and Deka (2014).

8.5   ADAPTIVE NEURO-FUZZY INFERENCE SYSTEM (ANFIS)

ANFIS is a soft computing technique designed by Jang (1993), and has been used in complex system control, modelling and parameter estimation (Şahin & Erol, 2017). In addition, ANFIS is a hybrid system that combines the characteristics of fuzzy logic with neural networks. The neural networks are supervised learning algorithms that predict future values using historical data. On the other hand, fuzzy logic uses the control signal produced from firing the rule base. The generated rule is based on historical data and is completely random. However, the neural network selects the rule base in the ANFIS method through BP and approximates the non-linear system setup of the IF–THEN rules (fuzzy logic) incorporated into the model technique. This is done to enhance the applicability and performance of the ANFIS technique, making it a universal estimator (Mathur et al., 2016). Furthermore, neural networks learn from the data, though it is difficult to comprehend the weight and neurons

TABLE 8.2
Advantages and Disadvantages of SVM

Advantages	Disadvantages
SRM principle provides SVM with the desirable property to maximize the margin. Hence, the generalization ability does not deteriorate and is capable of forecasting unseen data scenario (Raghavendra & Deka, 2014).	Poor model extrapolation occurs when there is inconsistency in the previous data, since the model is dependent on the past record as support vectors (Raghavendra & Deka, 2014).
SVM input vectors are very versatile. Thus, important variables like wind speed, temperature and relative humidity can easily be incorporated into the model (Moghaddamnia et al., 2008).	SVM provides point predictions, and it is not suited for probabilistic forecasting (Raghavendra & Deka, 2014).
SVM is capable of identifying and incorporating support vectors during the training phase, thus preventing the model from being influenced by non-support vectors. They enable the model to perform effectively in noisy environments (Han et al., 2007).	SVM model is difficult to understand due to the intrinsic complexity required in mapping non-linear input space into a high-dimensional feature space (Tripathi et al., 2006).
SVMs have the ability to reconstruct historical events in order to enhance future predictions by using lessons learned in the past (Han et al., 2007).	The selection of appropriate hyper parameters and kernel functions is based on trial and error (Ccoicca, 2013).
SVM is capable of providing reliable and robust classification results. Therefore, they can assist in the efficient evaluation of relevant information (Raghavendra & Deka, 2014).	SVM models become complex and expensive when dealing with large datasets (Yu et al., 2006).

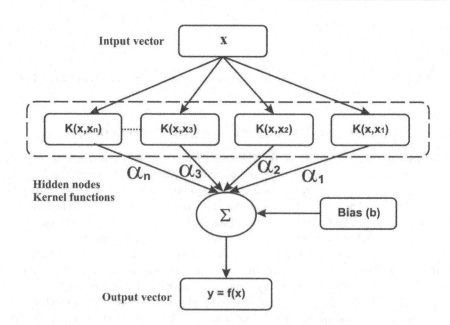

FIGURE 8.4 Architecture of SVM algorithm. Adapted from Nourani et al. (2018).

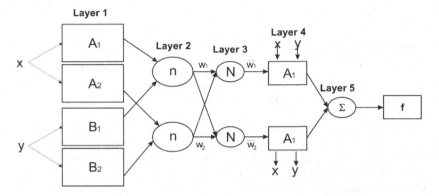

FIGURE 8.5   Structure of ANFIS. Adapted from Nourani et al. (2018).

associated with the data. Fuzzy logic cannot learn directly from data, but it is simple to understand since it uses linguistic terms rather than mathematical and IF–THEN structures. The fuzzy set in which linguistic variables are included is an extension of a "crisp" where an element could have full or no membership. On the other hand, a fuzzy set enables partial membership, meaning that an element might partially belong to more than one set (Mathur et al., 2016). In other words, this technique is a rule-based fuzzy logic model in which the rules are established during the training phase of the model (Ghiasi et al., 2016). ANFIS builds a fuzzy inference system based on the training examples. The two most frequently used fuzzy inference systems are the Sugeno and Mamdani systems. The distinction between the two fuzzy inference systems is that the former is either linear or constant while the latter is non-linear (Şahin & Erol, 2017).

In general, ANFIS is a FFNN structure comprised of neuro-fuzzy system components at each layer. The ANFIS model utilizes the fuzzy Sugeno model within an adaptive system framework to promote learning and adaptation. The framework enables more systematic ANFIS modelling, making it less reliant on expert knowledge. Additionally, the ANFIS structure is illustrated in Figure 8.5, Adapted from Nourani et al. (2018). The advantages and drawbacks of the ANFIS techniques are described in Table 8.3.

8.6 MULTILINEAR REGRESSION (MLR)

Multilinear regression analysis, also referred to as multiple regression, is a statistical method used in investigating and analysing the relationship between variables. It is a statistical method that involves constructing mathematical equations that relate the variable to independent variables or collection of predictors (Amid & Gundoshmian, 2017). In MLR, the variable to be modelled is known as the dependent variable, whereas the input data are known as the explanatory or independent variable. MLR is divided into linear and non-linear regression. Furthermore, linear regression model is used to predict the values of a dependent variable using a linear relationship with one or more predictors (Abba et al., 2017). The advantages and drawbacks of MLR are described in Table 8.4.

TABLE 8.3
Advantages and Drawbacks of ANFIS

Advantages	Drawbacks
In ANFIS technique, training of the models is independent of expert knowledge (Srisaeng et al., 2015).	ANFIS technique is sensitive to initial number of fuzzy rules (number of clusters) (Salleh et al., 2017).
ANFIS is capable of using both numerical and linguistic knowledge. Additionally, it applies the capabilities of ANN to categorize data and identify patterns (Srisaeng et al., 2015).	Computational complexity increases as the number of fuzzy rules increases (Salleh et al., 2017).
In comparison to ANN, ANFIS model is easier to use and creates less memory errors (Şahin & Erol, 2017).	
ANFIS approach has fast learning capacity, good adaptation capability and non-linear ability (Şahin & Erol, 2017).	

TABLE 8.4
Advantages and Drawbacks of MLR

Advantages	Drawbacks
MLR can detect outliers and anomalies very effectively (Rahman et al., 2012).	The problem of over-fitting is very prevalent in MLR (Rahman et al., 2012).
MLR is a technique for analysing the relationship between two or more regressors and a response variable (Rahman et al., 2012).	The precision decreases as the linearity of the dataset decreases (Rahman et al., 2012).
This approach is beneficial when attempting to account for confounding variables in observational research (Rahman et al., 2012).	

8.7   ERROR ENSEMBLE LEARNING APPROACH

Ensemble learning is a ML technique used in improving the final performance of prediction through the integration of multiple homogenous or heterogeneous models (Baba et al., 2015). This method gives precise results compared to single models when solving the same problem. Ensemble techniques have been used in different areas such as regression and time series problems, clustering and classification and web ranking algorithm (Kazienko et al., 2013). The fundamental element of ensemble learning is the base learner, which is generated using a base learning algorithm. The general formula for ensemble learning is presented in Equation 8.3 (Kazienko et al., 2013).

$$P_e(x) = \sum_{i=1}^{n} p_i(x) \tag{8.3}$$

where $p(x)$ and $P_e(x)$ represent single predictor and ensemble with n total number of predictors.

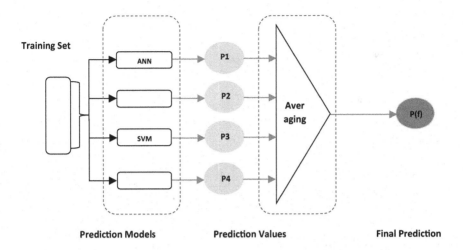

FIGURE 8.6 Structure of SAE.

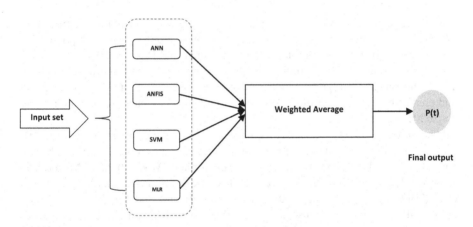

FIGURE 8.7 Structure of WAE.

Furthermore, ensemble learning is categorized into simple averaging ensemble (SAE), weighted averaging ensemble (WAE) and non-linear neural network ensemble (NNE). The structures of SAE, WAE and NNE are presented in Figures 8.6, 8.7 and 8.8, respectively.

8.8 DEVELOPMENT OF ANN, SVM, ANFIS AND MLR FOR PHYTOREMEDIATION OF WASTEWATER

This subsection presents recent applications of ANN, SVM, ANFIS and MLR in phytoremediation of wastewater. From previous work, it was discovered that ML techniques (ANN, SVM and ANFIS) have gained popularity in modelling complex

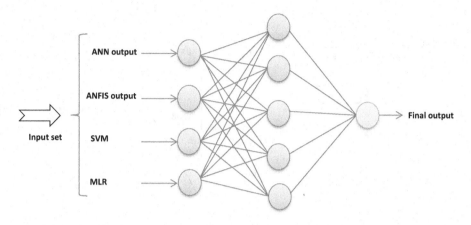

FIGURE 8.8   Structure of NNE.

wastewater treatment systems owing to their excellent performance in predicting WQ parameters involving large number of variables. Additionally, it is possible to use different ML models, like in the case of Kumar and Deswal (2020b) who used LR, ANN, RF and M5P to predict phosphorous removal using four aquatic plants. Thus, Nadiri et al. (2018) reported that the application of different ML for forecasting and predicting of water parameters would allow robust comparison study than using single ML model.

The study by Guo et al. (2015) was restricted to the applications of two ML models (ANN and SVM) and three performance criteria in forecasting the TN concentration of effluent from wastewater treatment plant (WWTP) using MATLAB® software. The data was normalized to 1 and −1, while Darajeh et al. (2017), Elfanssi et al. (2017) and Parveen et al. (2019) employed only one model in their study. Titah et al. (2018) employed response surface methodology (RSM) and ANN approaches in studying the phytoremediation of arsenic uptake by *Ludwigia octovalvis*. Furthermore, Nourani et al. (2018) applied four ML models (FFNN, SVM, MLR and ANFIS methods) in predicting the water quality indicators (TN, BOD and COD) of Nicosia WWTP. It was observed that ANFIS model outperforms all other models owing to the fuzzy concept capacity to deal with process uncertainty.

Vo et al. (2019) and Kumar et al. (2019) explored different regression models in their research. In contrast, Farzi and Borghei (2019) and Kumar et al. (2021) combined regression and ML models. Furthermore, Kumar and Deswal (2020a) conducted a comparative study on random forest (RF), M5P and ANN modelling techniques in predicting the reduction of phosphorous from rice mill wastewater. The trained data were divided into 70% and 30% for training and testing the model. The modelling findings suggest that ANN outperforms the M5P tree and RF modelling, while Kumar and Deswal (2020b) applied three statistical parameters (RMSE, R^2 and MAE) to analyse the model accuracy of linear regression (LR), ANN and M5P.

8.9   CONCLUSION

This chapter discussed the development of ANN, SVM, ANFIS, MLR and error ensemble learning approach for phytoremediation of wastewater. The aforementioned literature showed that two or more performance criteria are used in evaluating the accuracy of the models. R, R^2 and RMSE are the most frequently used performance criteria for the evaluation of the selected models. Furthermore, feature selection and data normalization process are critical at the first stage of model development. This may negatively affect the prediction accuracy of the data. Another factor that influences the performance and accuracy of the model is the volume of the data and the choice of the input variables. Hence, we recommend using a stepwise model-building approach or an analytical method like the cross-correlation methodology to choose input variables based on historical data. In this vein, available data on phytoremediation of wastewater is largely based on empirical studies. Therefore, we propose the application of promising ML tools (ANN, SVM and ANFIS), linear model (MLR) and four performance criteria (R, R^2, MSE and RMSE) for predicting and evaluating WQ data obtained from IoT devices owing to the fact that more research is required to develop promising models that would enhance WQ monitoring in phytoremediation of wastewater.

REFERENCES

Abba, S. I., Hadi, J. S., & Abdullahi, J. (2017). River water modelling prediction using multi-linear regression, artificial neural network, and adaptive neuro-fuzzy inference system techniques. *9th International Conference on Theory and Application of Soft Computing, Computing with 9th International Conference on Theory and Application of Soft Computing, Computing with Words and Perception, ICSCCW 2017, 22–23 August 2017, Budapest, Hungary Words*, *120*, 75–82. https://doi.org/10.1016/j.procs.2017.11.212

Abiodun, O. I., Jantan, A., Omolara, A. E., Dada, K. V., Mohamed, N. A., & Arshad, H. (2018). State-of-the-art in artificial neural network applications: A survey. *Heliyon*, *4*, 1–41. https://doi.org/10.1016/j.heliyon.2018.e00938

Abrahart, R. J., Anctil, F., Coulibaly, P., Dawson, C. W., Mount, N. J., See, L. M., Shamseldin, A. Y., Solomatine, D. P., Toth, E., & Wilby, R. L. (2012). Two decades of anarchy? Emerging themes and outstanding challenges for neural network river forecasting. *Progress in Physical Geography*, *36*(4), 480–513. https://doi.org/10.1177/0309133312444943

Adeyemo, J., Oyebode, O., & Stretch, D. (2018). River Flow Forecasting Using an Improved Artificial Neural Network. In A. A. Tantar, E. Tantar, M. Emmerich, P. Legrand, L. Alboaie, & H. Luchian (Eds.), *A Bridge between Probability, Set Oriented Numerics, and Evolutionary Computation VI* (Vol. 674, pp. 179–193). Springer, Cham. https://doi.org/10.1007/978-3-319-69710-9_13

Al Aani, S., Bonny, T., Hasan, S. W., & Hilal, N. (2019). Can machine language and artificial intelligence revolutionize process automation for water treatment and desalination? *Desalination*, *458*(February), 84–96. https://doi.org/10.1016/j.desal.2019.02.005

Amid, S., & Gundoshmian, T. M. (2017). Prediction of output energies for broiler production using linear regression, ANN (MLP, RBF), and ANFIS models. *Enivironmental Progress & Sustainable Energy*, *36*(2), 577–585. https://doi.org/10.1002/ep

Baba, N. M., Makhtar, M., Abdullah, S., & Awang, M. K. (2015). Current issues in ensemble methods and its applications. *Journal of Theoretical and Applied Information Technology*, *81*(2), 266–276.

Bagheri, M., Akbari, A., & Mirbagheri, S. A. (2019). Advanced control of membrane fouling in filtration systems using artificial intelligence and machine learning techniques: A critical review. *Process Safety and Environmental Protection*, *123*, 229–252. https://doi. org/10.1016/j.psep.2019.01.013

Basheer, I. A., & Hajmeer, M. (2000). Artificial neural networks: Fundamentals, computing, design, and application. *Journal of Microbiological Methods*, *43*(1), 3–31. https://doi. org/10.1016/S0167-7012(00)00201-3

Bennett, C., Stewart, R. A., & Beal, C. D. (2013). ANN-based residential water end-use demand forecasting model. *Expert Systems with Applications*, *40*(4), 1014–1023. https:// doi.org/10.1016/j.eswa.2012.08.012

Ccoicca, Y. J. (2013). Applications of support vector machines in the exploratory phase of petroleum and natural gas: A survey. *International Journal of Engineering & Technology*, *2*(2), 113. https://doi.org/10.14419/ijet.v2i2.834

Darajeh, N., Idris, A., Reza, H., Masoumi, F., Nourani, A., & Rezania, S. (2017). Phytoremediation of palm oil mill secondary effluent (POMSE) by Chrysopogon zizanioides (L.) using artificial neural networks Negisa. *International Journal of Phytoremediation*, *19*(5), 413–424. https://doi.org/10.1080/15226514.2016.1244159

Dibike, Y. B., Velickov, S., & Solomatine, D. (2000). Support Vector Machines: Review and Applications in Civil Engineering. *Proceedings of the 2nd Joint Workshop on Application of AI in Civil Engineering March 2000*, Cottbus, Germany.

Dinu, C., Drobot, R., Pricop, C., & Blidaru, T. V. (2017). Flash-flood modelling with artificial neural. *Mathematical Modelling in Civil Engineering*, *13*(3), 10–20. https://doi. org/10.1515/mmce

Elfanssi, S., Ouazzani, N., Latrach, L., & Hejjaj, A. (2017). Phytoremediation of domestic wastewater using a hybrid constructed wetlands in mountainous rural area. *International Journal of Phytoremediation*, *20*(June). https://doi.org/10.1080/15226514.2017.1337067

Elkiran, G., Nourani, V., & Abba, S. I. (2019). Multi-step ahead modelling of river water quality parameters using ensemble artificial intelligence-based approach. *Journal of Hydrology*, *577*(July), 123962. https://doi.org/10.1016/j.jhydrol.2019.123962

Esfe, M. H. (2017). Designing a neural network for predicting the heat transfer and pressure drop characteristics of Ag/water nanofluids in a heat exchanger. *Applied Thermal Engineering*, *126*, 559–565. https://doi.org/10.1016/j.applthermaleng.2017.06.046

Fan, M., Hu, J., Cao, R., Ruan, W., & Wei, X. (2018). A review on experimental design for pollutants removal in water treatment with the aid of artificial intelligence. *Chemosphere*, *200*, 330–343. https://doi.org/10.1016/j.chemosphere.2018.02.111

Farzi, A., & Borghei, S. M. (2019). Modeling of salt phytoremediation in constructed wetlands containing halophytic plants using Artificial Neural Network. *Journal of Environmental Science Studies*, *4*(2), 1373–1380.

Gaya, M. S., Abdul Wahaba, N., Sama, Y. M., & Samsudin, S. I. (2014). ANFIS modelling of carbon and nitrogen removal in domestic wastewater treatment plant. *Jurnal Teknologi (Sciences & Engineering)*, *67*(5), 29–34. https://doi.org/https://doi.org/10.11113/jt.v67.2839

Gayaa, M. S., Abdul Wahaba, N., Sama, Y. M., & Samsudin, S. I. (2014). ANFIS modelling of carbon and nitrogen removal in domestic wastewater treatment plant. *Jurnal Teknologi*, *67*(5), 439–446.

Ghiasi, M. M., Arabloo, M., Mohammadi, A. H., & Barghi, T. (2016). Application of ANFIS soft computing technique in modeling the CO2 capture with MEA, DEA, and TEA aqueous solutions. *International Journal of Greenhouse Gas Control*, *49*, 47–54. https:// doi.org/10.1016/j.ijggc.2016.02.015

Guo, H., Jeong, K., Lim, J., Jo, J., Kim, Y. M., Park, J.-P., Kim, J. H., & Cho, K. H. (2015). Prediction of effluent concentration in a wastewater treatment plant using machine learning models. *Journal of Environmental Sciences (China)*, *32*(April), 90–101. https://doi. org/10.1016/j.jes.2015.01.007

Hamed, M. M., Khalafallah, M. G., & Hassanien, E. A. (2004). Prediction of wastewater treatment plant performance using artificial neural networks. *Environmental Modelling & Software, 19*(10), 919–928. https://doi.org/10.1016/j.envsoft.2003.10.005

Han, D., Chan, L., & Zhu, N. (2007). Flood forecasting using support vector machines. *Journal of Hydroinformatics, 9*(4), 267–276. https://doi.org/10.2166/hydro.2007.027

Hayder, G., Ramli, M. Z., Malek, M. A., Khamis, A., & Hilmin, N. M. (2014). Prediction model development for petroleum refinery wastewater treatment. *Journal of Water Process Engineering, 4*, 1–5. https://doi.org/10.1016/j.jwpe.2014.08.006

Haykin, S. (1999). Neural networks: A comprehensive foundation by Simon Haykin. *The Knowledge Engineering Review, 13*(4), 409–412. https://doi.org/10.1017/S026988899 8214044

Jang, J. R. (1993). ANFIS: Adaptive-network-based fuzzy inference system. *IEEE Transactions on Systems, Man, and Cybernetics, 23*(3), 665–685.

Kazienko, P., Lughofer, E., & Trawiński, B. (2013). Hybrid and ensemble methods in machine learning. *Journal of Universal Computer Science, 19*(4), 457–461.

Khan, M. S., & Coulibaly, P. (2006). Application of support vector machine in lake water level prediction. *Journal of Hydrologic Engineering, 11*(3), 199–205. https://doi.org/10.1061/(ASCE)1084-0699(2006)11:3(199)

Kumar, S., & Deswal, S. (2020a). Estimation of phosphorus reduction from wastewater by artificial neural estimation of phosphorus reduction from wastewater by artificial neural network, random forest and M5P model tree approaches. *Pollution, 6*(2), 417–428. https://doi.org/10.22059/poll.2020.293086.717

Kumar, S., & Deswal, S. (2020b). Phytoremediation capabilities of Salvinia molesta, water hyacinth, water lettuce, and duckweed to reduce phosphorus in rice mill wastewater. *International Journal of Phytoremediation, 22*(11), 1097–1109. https://doi.org/10.1080/15226514.2020.1731729

Kumar, V., Kumar, P., Singh, J., & Kumar, P. (2021). Use of sugar mill wastewater for Agaricus bisporus cultivation: prediction models for trace metal uptake and health risk assessment. *Environmental Science and Pollution Research.* https://doi.org/10.1007/s11356-021-12488-7

Kumar, V., Singh, J., & Kumar, P. (2019). Heavy metal uptake by water lettuce (Pistia stratiotes L.) from paper mill effluent (PME): Experimental and prediction modeling studies. *Environmental Science and Pollution Research, 26*, 14400–14413.

Maier, H. R., & Dandy, G. C. (2000). Neural networks for the prediction and forecasting of water resources variables: A review of modelling issues and applications. *Environmental Modelling & Software, 15*(1), 101–124. https://doi.org/10.1016/S1364-8152(99)00007-9

Mathur, N., Glesk, I., & Buis, A. (2016). Comparison of adaptive neuro-fuzzy inference system (ANFIS) and Gaussian processes for machine learning (GPML) algorithms for the prediction of skin temperature in lower limb prostheses. *Medical Engineering and Physics, 38*(10), 1083–1089. https://doi.org/10.1016/j.medengphy.2016.07.003

Moghaddamnia, A., Ghafari, M., Piri, J., & Han, D. (2008). Evaporation estimation using support vector machines technique. *World Academy of Science, Engineering and Technology, 43*, 14–22.

Nadiri, A. A., Shokri, S., Tsai, F. T.-C., & Moghaddam, A. A. (2018). Prediction of effluent quality parameters of a wastewater treatment plant using a supervised committee fuzzy logic model. *Journal of Cleaner Production, 180*(February), 539–549. https://doi.org/10.1016/j.jclepro.2018.01.139

Nourani, V., Elkiran, G., & Abba, S. I. (2018). Wastewater treatment plant performance analysis using artificial intelligence - An ensemble approach. *Water Science and Technology, 78*(10), 2064–2076. https://doi.org/10.2166/wst.2018.477

Oyebode, O., & Stretch, D. (2019). Neural network modeling of hydrological systems: A review of implementation techniques. *Natural Resource Modeling, 32*(1), 1–14. https://doi.org/10.1111/nrm.12189

Pang, J., Yang, S., He, L., Chen, Y., & Nanqi, R. (2019). Intelligent control/operational strategies in WWTPs through an integrated Q-Learning Algorithm with ASM2d-guided reward. *Water,* *11*(927), 1–18.

Parveen, N., Zaidi, S., & Danish, M. (2019). Groundwater for sustainable development support vector regression (SVR)-based adsorption model for Ni (II) ions removal. *Groundwater for Sustainable Development,* *9*(May), 100232. https://doi.org/10.1016/j.gsd.2019.100232

Raghavendra, S., & Deka, P. C. (2014). Support vector machine applications in the field of hydrology: A review. *Applied Soft Computing Journal,* *19*, 372–386. https://doi.org/10.1016/j.asoc.2014.02.002

Rahman, S. M. A. K., Sathik, M. M., & Kannan, K. S. (2012). Multiple linear regression models in outlier detection. *International Journal of Research in Computer Science,* *2*(2), 23–28. https://doi.org/10.7815/ijorcs.22.2012.018

Rajaee, T., Ebrahimi, H., & Nourani, V. (2019). A review of the artificial intelligence methods in groundwater level modeling. *Journal of Hydrology,* *572*(May 2018), 336–351. https://doi.org/10.1016/j.jhydrol.2018.12.037

Şahin, M., & Erol, R. (2017). A comparative study of neural networks and ANFIS for forecasting attendance rate of soccer games. *Mathematical and Computational Applications,* *22*(43), 1–12. https://doi.org/10.3390/mca22040043

Salleh, M. N. M., Talpur, N., & Hussain, K. (2017). Adaptive neuro-fuzzy inference system: Overview, strengths, limitations, and solutions. *Lecture Notes in Computer Science (Including Subseries Lecture Notes in Artificial Intelligence and Lecture Notes in Bioinformatics),* *10387 LNCS*(August), 527–535. https://doi.org/10.1007/978-3-319-61845-6_52

Senthil Kumar, A. R., Goyal, M. K., Ojha, C. S. P., Singh, R. D., & Swamee, P. K. (2013). Application of artificial neural network, fuzzylogic and decision tree algorithms for modelling of streamflow at Kasol in India. *Water Science and Technology,* *68*(12), 2521–2526. https://doi.org/10.2166/wst.2013.491

Sharghi, E., Nourani, V., Najafi, H., & Molajou, A. (2018). Emotional ANN (EANN) and Wavelet-ANN (WANN) approaches for markovian and seasonal based modeling of rainfall-runoff process. *Water Resources Management,* *32*(10), 3441–3456. https://doi.org/10.1007/s11269-018-2000-y

Shi, S., & Xu, G. (2018). Novel performance prediction model of a biofilm system treating domestic wastewater based on stacked denoising auto-encoders deep learning network. *Chemical Engineering Journal,* *347*(February), 280–290. https://doi.org/10.1016/j.cej.2018.04.087

Srisaeng, P., Baxter, G. S., & Wild, G. (2015). An adaptive neuro-fuzzy inference system for forecasting Australia's domestic low cost carrier passenger demand. *Aviation,* *19*(3), 150–163. https://doi.org/10.3846/16487788.2015.1104806

Titah, H. S., Izuan, M., Bin, E., Rozaimah, S., Abdullah, S., Hasan, H. A., Idris, M., & Anuar, N. (2018). Statistical optimization of the phytoremediation of arsenic by Ludwigia octovalvis- in a pilot reed bed using response surface methodology (RSM) versus an artificial neural network (ANN). *International Journal of Phytoremediation,* *20*(7). https://doi.org/10.1080/15226514.2017.1413337

Tripathi, S., Srinivas, V. V., & Nanjundiah, R. S. (2006). Downscaling of precipitation for climate change scenarios: A support vector machine approach. *Journal of Hydrology,* *330*, 621–640. https://doi.org/10.1016/j.jhydrol.2006.04.030

Turunen, V., Sorvari, J., & Mikola, A. (2018). A decision support tool for selecting the optimal sewage sludge treatment. *Chemosphere,* *193*, 521–529. https://doi.org/10.1016/j.chemosphere.2017.11.052

Vapnik, V. (1998). *Statistical Learning Theory.* Wiley-Interscience, New York.

Vo, H. N.., Koottatep, T., Chapagain, S.., Panuvatvanich, A., Polprasert, C., Nguyen, T. M. H., Chaiwong, C., & Nguyen, N. L. (2019). Removal and monitoring acetaminophen-contaminated hospital wastewater by vertical flow constructed wetland and peroxidase enzymes. *Journal of Environmental Management, 250*(September), 109526. https://doi.org/10.1016/j.jenvman.2019.109526

Wernick, M. N., Yang, Y., Brankov, J. G., Yourganov, G., & Strother, S. C. (2010, July). Drawing conclusions from medical images. *IEEE Signal Processing Magazine, July*, 25–38. https://doi.org/1053-5888/10/$26.00©2010IEEE

Xu, Y., Wang, Z., Jiang, Y., Yang, Y., & Wang, F. (2019). Small-world network analysis on fault propagation characteristics of water networks in eco-industrial parks. *Resources, Conservation and Recycling, 149*(May), 343–351. https://doi.org/10.1016/j.resconrec.2019.05.040

Yu, P. S., Chen, S. T., & Chang, I. F. (2006). Support vector regression for real-time flood stage forecasting. *Journal of Hydrology, 328*(3–4), 704–716. https://doi.org/10.1016/j.jhydrol.2006.01.021

Zhao, L., Dai, T., Qiao, Z., Sun, P., Hao, J., & Yang, Y. (2020). Application of artificial intelligence to wastewater treatment: A bibliometric analysis and systematic review of technology, economy, management, and wastewater reuse. *Process Safety and Environmental Protection, 133*(92), 169–182. https://doi.org/10.1016/j.psep.2019.11.014

Zhu, J., Kang, L., & Anderson, P. R. (2018). Predicting in fluent biochemical oxygen demand: Balancing energy demand and risk management. *Water Research, 128*, 304–313. https://doi.org/10.1016/j.watres.2017.10.053

9 Case Study
Monitoring and Evaluation of Phytoremediation System Using Internet of Things (IoT) and Machine Learning Techniques

9.1 INTRODUCTION

Over the past century, there has been increasing efforts worldwide devoted to the development of water quality models (Whitehead et al., 2019). Monitoring activities can help in understanding, protecting and improving aquatic habitat. Water quality data analysis helps in quantifying environmental changes and develop best management practices. Therefore, water quality monitoring network is a key element for managing and protecting water environment as it captures information about the state of water systems (Jiang et al., 2020). Additionally, smart water quality monitoring is regarded as the future water quality monitoring technology that enhances efficient data collection, communication, data analysis and early warning (Dong et al., 2015). Data transferred between sensors and the core network such as internet of things (IoT) technology generates huge data more frequently and also offloads computation and storage to the cloud (Saravanan et al., 2018). For example, IoT environmental sensor is a valuable tool for monitoring and modelling environmental phenomena (Edmondson et al., 2018; Mustafa et al., 2021c). Additionally, sensors are connected to a network of different chipboards such as the Arduino microprocessors, ESP32, ESP8266 and ARM-based Raspberry pi microcomputer in order to collect data that can be viewed through a computer and interpreted using Artificial Intelligence (AI) tools (Chowdury et al., 2019; Mustafa et al., 2021c). Furthermore, a lot of research has been reported using IoT sensors for measuring water quality parameters (Saravanan et al., 2018). Nevertheless, little or no study has been done to integrate the application of smart Arduino IoT system and machine learning techniques for monitoring and prediction of water quality parameters in phytoremediation of domestic wastewater using *Salvinia molesta* plants. Thus, this study aims to bridge this gap.

Moreover, the efficiency of *S. molesta* plants in phytoremediation of domestic wastewater has been reported (Hayder & Mustafa, 2021; Mustafa et al., 2021a;

DOI: 10.1201/9781003359586-9

Mustafa & Hayder, 2020a, 2020b, 2021a, 2021b, 2021c, 2021d). Mustafa and Hayder (2021c) investigated the best conditions for efficient applications of water lettuce, giant salvinia and water hyacinth in improving the quality of low-strength domestic wastewater based on varying retention times. Hendriarianti and Soetedjo (2021) employed Thingspeak as an IoT technology to monitor the phytoremediation of wastewater processes. In their proposed real-time monitoring system, several phytoremediation models were implemented on embedded hardware and connected to the Thingspeak IoT platform. Arismendy et al. (2020) applied intelligent system and multilayer perceptron neural network for predictive analysis of the chemical oxygen demand (COD) in industrial wastewater treatment process. Therefore, this research paper proposed the integration of IoT intelligent system, machine learning tools and classical method using the data collected from the phytoremediation process in modelling and prediction of the turbidity water quality parameter. This research would provide support in decision making related to the application of phytoremediation process in sewage treatment plants.

9.2 DEVELOPMENT OF INTERNET OF THINGS (IoT)-BASED *SALVINIA MOLESTA* PLANTS IN WASTEWATER TREATMENT

The Arduino Board used in this research consists of sensor nodes made up of PIC microcontroller, liquid crystal display (LCD) and sim card, which was used in reading the turbidity of the influent and effluent water samples. The Arduino Board UNO was connected to the internet through an external Wi-Fi module while the sensor nodes were inserted in the water samples. The data from the sensors was displayed on the liquid crystal display and directly to the computer. The connection diagram for an IoT system is shown in Figure 9.1.

9.2.1 COLLECTION OF DATA

The Arduino IoT device was used to monitor the phytoremediation of the influent wastewater cultivated with fresh *S. molesta* plants. The 2211 water quality parameter data was generated from the IoT monitoring system. The generated IoT data was pre-processed using descriptive statistic, sensitivity analysis and normalization in order to clean and filter the raw data to obtain a uniform input and output data. Figure 9.2 depicts the data collection process of the influent and effluent water samples.

9.2.2 MODELLING AND PREDICTION OF THE TURBIDITY (TURBT)

The programming execution was performed in MATLAB® R2020a (9.8.0.1417392) based on the 2211 data extracted from the Arduino IoT device. Machine learning techniques such as artificial neural network (ANN), support vector machine (SVM) and adaptive neuro-fuzzy inference system (ANFIS), classical method and multilinear regression (MLR) were employed in the modelling and prediction of the turbidity parameter. Two models (M1 and M2) were generated for the evaluation and the model validation was conducted using four performance criteria. The data from

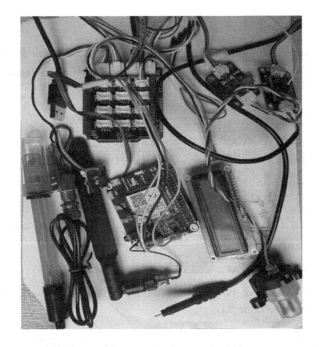

FIGURE 9.1 Connection diagram of an IoT system.

(a)   (b)

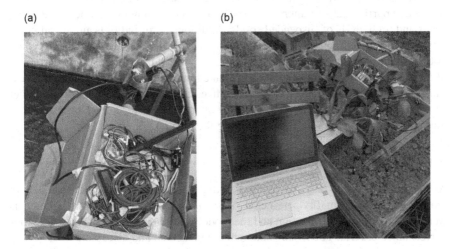

FIGURE 9.2 (a) Influent data collection and (b) effluent data collection.

the influent water was used as the input variables, while the data obtained from the effluent water samples served as the output variables. Furthermore, coefficient of determination (R^2), mean square error (MSE), correlation coefficient (R) and root mean square error (RMSE) were used to calculate and compare the performance of each model.

TABLE 9.1
Parameter Used (Influent (Raw) and Effluent (Treated) Samples)

Parameters	Influents Parameter	Effluent Parameter
Raw turbidity	TURBr	TURBt
Treated turbidity	TURBt	
Raw total dissolved solid	TDSr	
Treated total dissolved solid	TDSt	
Raw oxidation–reduction potential	ORPr	
Treated oxidation–reduction potential	ORPt	
Raw temperature	TEMPr	
Treated temperature	TEMPt	

9.2.3 WATER QUALITY PARAMETERS USED

The water quality parameters used in the modelling and evaluation of the turbidity (TURBt) are given in Table 9.1.

9.2.4 INFLUENT AND EFFLUENT CONCENTRATION OF TURBIDITY

In most cases, high turbidity in water can cause increased water temperatures because suspended particles absorb more heat and reduce the amount of light penetrating the water and high levels of suspended solids. Additionally, turbidity is an optical property that does not directly reflect the quantity or type of suspended solids (Kitt et al., 2005). The time-series and boxplots for influent and effluent turbidity concentrations of the *S. molesta* treatment system are presented in Figure 9.3. Boxplot is a graphical illustration of statistical datasets based on five-number summary. The five-number summary describes the first quartile, third quartile, median, minimum and maximum. EViews software was used in plotting the boxplot graphs.

9.3 RESULTS AND DISCUSSION

As stated earlier, this research applied ANN, SVM, ANFIS and MLR models in predicting the water quality parameters of *S. molesta* treatment system. Two models were generated for the evaluation and the model validation was conducted using four performance criteria. The results obtained are presented and discussed.

9.3.1 RESULTS OF TURBt (ANN, SVM, ANFIS, AND MLR)

In the ANN simulation, the maximum number of MSE, iterations and learning rate were fixed as 0.0001, 10,000 and 0.01, respectively. The log sigmoid and purlin was observed to be the best activation functions for the hidden and output layers, respectively. Furthermore, in order to generate the optimal models, the appropriate number of hidden nodes must be selected to avoid over-fitting, whereas inadequate neurons

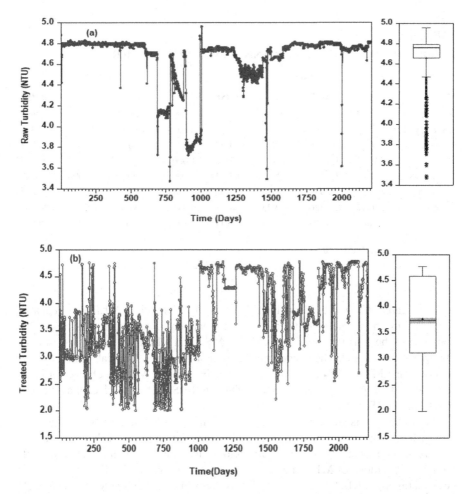

FIGURE 9.3 Time-series and boxplot of turbidity concentrations of the hydroponic tanks at the (a) influent and (b) effluent (treated).

can capture unsatisfactory information (Moosavi et al., 2012; Koutroumanidis et al., 2009). Additionally, the ideal number of nodes for determining the optimum number of hidden layers is $(n + 1)$, where n represents the number of inputs (Elkiran et al., 2019). Therefore, the range of hidden neurons in a typical three-layer ANN model for forecasting the turbidity effluent water sample (TURBt) was observed to be 5–21.

Furthermore, different membership functions were evaluated in the ANFIS modelling through trial-and-error method to obtain the ANFIS structure, with each structure formed by different epoch, iterations and membership functions. Additionally, different SVM models were used for all the input combinations. The determination of the optimal combination of C and g parameters is extremely necessary to achieve the best precision in generating the SVM model. Therefore, the procedure of grid check was then used to evaluate the optimal values (Granata et al., 2017). Table 9.2 represents the results of the TURBt prediction.

TABLE 9.2

TURBt Performance Criteria for ANN, SVM, ANFIS and MLR

	Training				Validation			
Models	R^2	MSE	R	RMSE	R^2	MSE	R	RMSE
ANN-M1	0.9997	0.1410	0.9999	0.3755	0.9998	0.0677	0.9999	0.2603
ANN-M2	0.9989	0.5790	0.9994	0.7609	0.9915	3.3591	0.9957	1.8328
SVM-M1	0.9999	0.0006	0.9999	0.0251	0.9997	0.1083	0.9999	0.3291
SVM-M2	0.9985	0.7684	0.9993	0.8766	1.0000	0.0770	1.0000	0.2775
ANFIS-M1*	0.9999	0.0001	0.9999	0.0071	1.0000	0.0001	1.0000	0.0109
ANFIS-M2	0.9953	2.3977	0.9977	1.5484	0.9858	5.5982	0.9929	2.3661
MLR-M1	0.9947	2.7320	0.9973	1.6529	0.9838	6.3788	0.9919	2.5256
MLR-M2	0.9889	5.7203	0.9944	2.3917	0.9660	13.3559	0.9829	3.6546

* Signifies the overall best model.

From Table 9.2, it is evident that ANN, SVM, ANFIS and MLR were capable of modelling the turbidity of the *S. molesta* treatment system; this was proved by considering the statistical indicators (R, R^2, MSE and RMSE). The sign (*) implies the overall best model. The visual investigation of the models indicated that M1 was superior than M2 for all the models. Additionally, ANFIS-M1 was observed to perform better than ANN, SVM and MLR. Thus, ANFIS-M1* was selected as the best overall in TURBt modelling.

Additionally, it is worth mentioning that all the data-driven models employed for the prediction of TURBt depicted reasonable accuracy in terms of the performance criteria of the models. However, a satisfactory outcome was obtained from the different combinations of M1 and M2. From the simulation of TURBt, M1 with four combinations (TURB, ORP, TDS and TEMP) proved to be the best for all the models in comparison with M2 (TURB, ORP, TDS) with three combinations. From the literature point of view, it can be deduced that TEMP was crucial in the prediction of the water quality parameters.

Figures 9.4 and 9.5 represent the time-series graphs between the observed and predicted TURBt concentrations by M1 and M2 for ANN, SVM, ANFIS and MLR. Further evaluation of the time-series findings demonstrates that ANN-M1, SVM-M1, ANFIS-M1 and MLR-M1 gave a satisfactory outcome, while an acceptable precision was obtained for both ANN-M2, SVM-M2, ANFIS-M2 and MLR-M2. The promising potential of the ANFIS model was not surprising, since it is an emerging non-linear hybrid model and has demonstrated more excellent predictive abilities in different fields (Ahmed et al., 2017; Alas et al., 2020; Gaya et al., 2014; Yaseen et al., 2018). Furthermore, the quantitative examination of ANFIS-M1 in terms of MSE = 0.0001 and RMSE = 0.0071 performed better than ANN, SVM and MLR. Similarly, this outcome demonstrated that the influence for each independent WQ model behaves differently for the same input combinations. Additionally, time-series graphs are valuable methods for representation of different statistical scenarios.

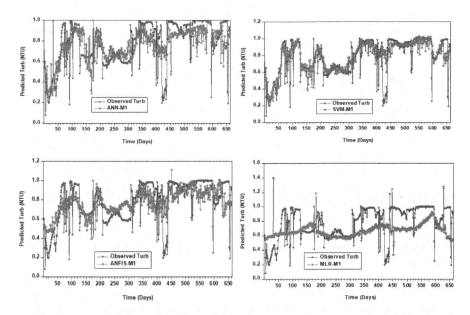

FIGURE 9.4   Turbidity time-series plot for ANN-M1, ANFIS-M1, SVM-M1 and MLR-M1 for validation phase.

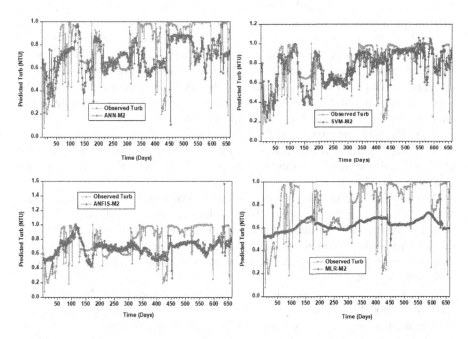

FIGURE 9.5   Turbidity time-series plot for ANN-M2, ANFIS-M2, SVM-M2 and MLR-M2 for validation phase.

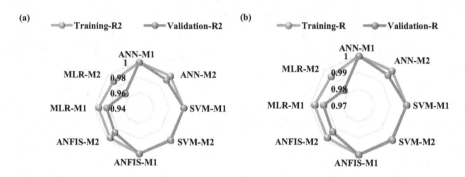

FIGURE 9.6   Radar chart for (a) goodness-of-fit and (b) correlation coefficient for turbidity.

Trends can easily be identified using time-series graphs and can be used for future forecasting.

It was evident that the time-series of the models vary from one combination to the other; this implies that the vibrational pattern of the M models depends on how the models capture the relationship between the observed and the target parameters. From the above figures, it can be justified that ANFIS-M1 captures the pattern of the time-series more than the other models, with the worst models being MLR-M2. The uniqueness of the goodness-of-fit could be indicated in the radar chart. Despite the over-fitting related to the goodness-of-fit, it was taken care of by the other performance criteria of the models. On the other hand, it is important to involve two or more performance criteria in order to come up with a justifiable argument.

Furthermore, radar chart is a type of graph used to compare three or more variables on a 2-dimensional plane. Radar charts can easily be used in depicting several variables without creating a clutter and they are viewed as better substitute for column graphs. Thus, radar charts were used to display the goodness-of-fit and R of the simulated turbidity model. Figure 9.6 shows the goodness-of-fit and R for turbidity. Additionally, MS Excel was used in plotting the radar charts.

According to Figure 9.6, the radar chart ranged from 0 to 1, with the best value approaching 1. It was observed that ANFIS-M1 is on the last spider web for both the goodness-of-fit and R. This proves the accuracy of the training and validation results presented in Table 9.2.

Additionally, boxplots are used to illustrate overall trends (range and other characteristics) of responses for a group. The boxplot illustration for M1 and M2 is presented in Figure 9.7. The understanding of the entire concept of boxplot is quite different with the evaluation criteria that displayed the unique digit of the number according to ranges. Figure 9.7 represents the boxplot graph of all the entire instances of the validation data.

From Figure 9.7, it is apparent that promising outcomes were observed, since the boxplot model gave precise information on how the values were equally distributed, making it easier to compare the distribution between the data. As a result, the variation was identical to the data spread but differs from the model's quantitative accuracy.

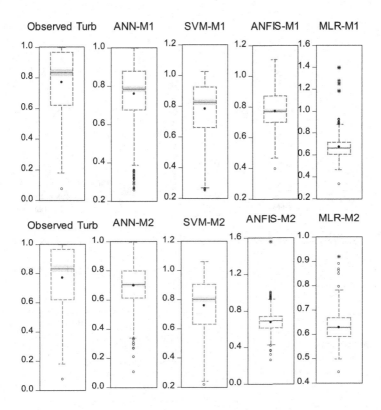

FIGURE 9.7 Boxplot for M1 and M2 in the validation phase for turbidity.

Despite the outliers indicated by other models, the overall results show the promising capability of ANFIS-M1 with regard to the spread of the graph. MLR generally provided the extent of outliers, followed by ANFIS-M2, ANN-M1 and ANN-M2. The outliers observed in the ANN models can be attributed to the research by Abba et al. (2017); Lee et al. (2018); Liu et al. (2013) and Pomeroy (2007), who reported that ANN models are known to inherit over-fitting during training which is probably the reason associated with the outlier. Additionally, the satisfactory performance obtained in the model combinations might be attributed to the direct relations established by the positive correlation between the observed turbidity (TURBt) and the other water quality parameters. These findings were similar to the results reported by Zhu and Heddam (2019).

Furthermore, a two-dimensional Taylor diagram was used to depict the comparison's evaluation for the best models. *R*-package was used in plotting the Taylor diagrams. The Taylor diagram summarizes and highlights statistical indices that included *R*, RMSE and the standard deviation between the observed and the predicted values (Ghorbani et al., 2018; Taylor, 2001). In addition, Taylor diagrams offer a means to graphically illustrate how exactly a pattern (or group of patterns) fits observations. The *R*, RMSE and standard deviations can be used to quantify the similarity between two patterns. These graphs are particularly helpful in evaluating

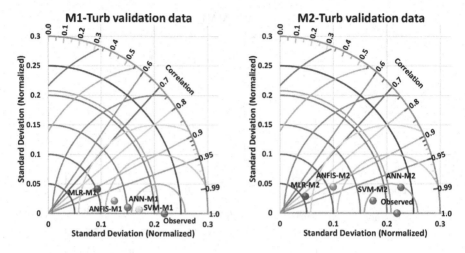

FIGURE 9.8 2-Dimensional Taylor diagram between the observed and predicted turbidity in the validation phase.

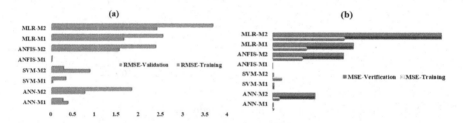

FIGURE 9.9 Error plot in both training and validation for (a) MSE and (b) RMSE.

various facets of complex models or calculating the relative ability of several different models (Taylor, 2001). Figure 9.8 represents the Taylor diagram between the observed and the predicted turbidity for M1 and M2. Additionally, Figures 9.9 and 9.10 represent the error plot in both training and validation, and the marginal correlation between the observed and predicted models in the validated phase. MS Excel was used in plotting the error plots.

From Figure 9.8, it is obvious that the simulated TURBt achieved improved fit using the ML models with the ranking order of ANFIS > SVM > ANN > MLR for M1 and ANN > SVM > ANFIS > MLR for M2. This outcome shows that ML techniques could capture the complex non-linear patterns between the WQ variables for both training and validation phase. Additionally, the numerical comparison of ML models with regard to RMSE depicted that ANFIS-M1 decreases the prediction accuracy of SVM-M1 and ANN-M1 by 31% and 24%, respectively (see Figure 9.9). Additionally, the ANFIS model led to the best fit with the observed data. This could be attributed to the ANFIS structure as it integrates the advantage of fuzzy reasoning and the self-learning ability of neural networks and thus gives a strong capability of eliminating noise (Rajaee et al., 2009). Furthermore, the results recorded in this

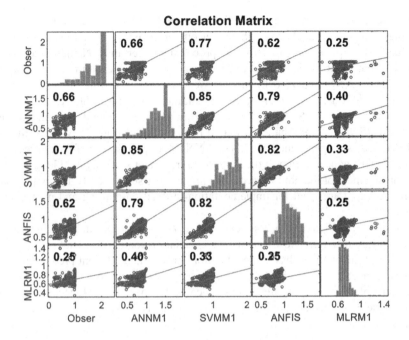

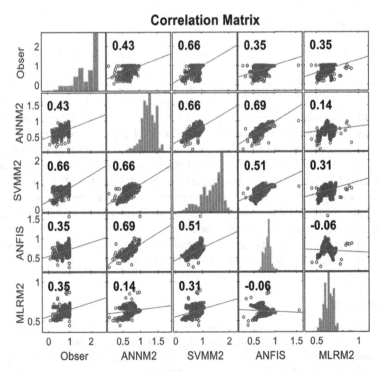

FIGURE 9.10 The marginal internal correlation between the observed and predicted models in the validation phase.

study are in agreement with the findings reported by Nourani et al. (2018). Their findings indicated that the ANFIS model outperforms all other models owing to the fuzzy concept capacity to deal with processing uncertainty. Nemati et al. (2015) reported that MLR model did not have a high degree of precision in estimating DO when MLR, ANN and ANFIS models were applied in estimating water quality parameters of Tai Po River, Hong Kong. Additionally, Al-Mukhtar and Al-Yaseen (2019) employed ANFIS, ANN and MLR neural networks in modelling and predicting the TDS and EC of Abu-Ziriq marsh in the south of Iraq. They found that the ANFIS models outperformed MLR and ANN. More precise and efficient prediction of the TURBt concentration in the *S. molesta* treatment system would lead to effective management of the phytoremediation process. As such, the ML techniques used in this research are ideal for implementation in decision-making and management processes.

As shown in Figure 9.10, the scatter plot, R, distribution and 1:1 lines of the observed turbidity were close in marginal correlation comparison with the predicted ANFIS and the other AI-based models. More precise and efficient prediction of the TURBt concentration in the *S. molesta* treatment system would lead to effective management of the phytoremediation process. As such, the ML techniques used in this research are ideal for implementation in decision-making and management processes.

9.4 CONCLUSION

This research proposed the applications of IoT device, machine learning models (ANN, SVM, ANFIS) and one classical method (MLR) in monitoring and modelling the turbidity reduction of the influent wastewater treatment using *S. molesta* plants. The descriptive statistical analysis and correlation matrix was applied to investigate the extent, degree and type of relation between the influent and effluent parameters. The correlation coefficient (R) was evaluated to determine the extent of linear interaction between the two variables. It is obvious that the simulated TURBt achieved improved fit using an AI-based model with the ranking order of ANFIS>SVM>ANN>MLR for M1 and ANN >SVM>ANFIS>MLR for M2. This outcome shows that AI could capture the complex non-linear patterns between the WQ variables for both calibration and verification. Furthermore, taking into account the outcome of the single model and variation in the models, it can be inferred that there is a need for further studies to develop a degree of consistency between the observed and predicted datasets in terms of the values.

REFERENCES

Abba, S. I., Hadi, J. S., & Abdullahi, J. (2017). River water modelling prediction using multi-linear regression, artificial neural network, and adaptive neuro-fuzzy inference system techniques. *9th International Conference on Theory and Application of Soft Computing, Computing with 9th International Conference on Theory and Application of Soft Computing, Computing with Words and Perception, ICSCCW 2017, 22–23 August 2017, Budapest, Hungary Words, 120*, 75–82. https://doi.org/10.1016/j.procs.2017.11.212

Ahmed, A. A. M., Mustakim, S., & Shah, A. (2017). Application of adaptive neuro-fuzzy inference system (ANFIS) to estimate the biochemical oxygen demand (BOD) of Surma River. *Journal of King Saud University - Engineering Sciences*, *29*(3), 237–243. https://doi.org/10.1016/j.jksues.2015.02.001

Al-Mukhtar, M., & Al-Yaseen, F. (2019). Modeling water quality parameters using data-driven models, a case study Abu-Ziriq marsh in south of Iraq. *Hydrology*, *6*(24), 17. https://doi.org/10.3390/hydrology6010024

Alas, M., Ali, S. I. A., Abulhadi, Y., & Abba, S. I. (2020). Experimental evaluation and modeling of polymer nanocomposite modified asphalt binder using ANN and ANFIS. *Journal of Materials in Civil Engineering*, *32*(10), 1–11. https://doi.org/10.1061/(ASCE)MT.1943-5533.0003404

Arismendy, L., Cárdenas, C., Gómez, D., Maturana, A., Mejía, R., & Quintero M., C. G. (2020). Intelligent system for the predictive analysis of an industrial wastewater treatment process. *Sustainability (Switzerland)*, *12*(16). https://doi.org/10.3390/SU12166348

Chowdury, M. S. U., Emran, T. B., Ghosh, S., Pathak, A., Alam, M. M., Absar, N., Andersson, K., & Hossain, M. S. (2019). IoT based real-time river water quality monitoring system. *Procedia Computer Science*, *155*, 161–168. https://doi.org/10.1016/j.procs.2019.08.025

Dong, J., Wang, G., Yan, H., Xu, J., & Zhang, X. (2015). A survey of smart water quality monitoring system. *Environmental Science and Pollution Research*, *22*(7), 4893–4906. https://doi.org/10.1007/s11356-014-4026-x

Edmondson, V., Cerny, M., Lim, M., Gledson, B., Lockley, S., & Woodward, J. (2018). A smart sewer asset information model to enable an 'Internet of Things' for operational wastewater management. *Automation in Construction*, *91*, 193–205. https://doi.org/10.1016/j.autcon.2018.03.003

Elkiran, G., Nourani, V., & Abba, S. I. (2019). Multi-step ahead modelling of river water quality parameters using ensemble artificial intelligence-based approach. *Journal of Hydrology*, *577*(April), 1–12. https://doi.org/10.1016/j.jhydrol.2019.123962

Gaya, M. S., Abdul Wahaba, N., Sama, Y. M., Samsudin, S. I. (2014). ANFIS modelling of carbon and nitrogen removal in domestic wastewater treatment plant. *Jurnal Teknologi*, *67*(5), 439–446.

Ghorbani, M. A., Deo, R. C., Yaseen, Z. M., Kashani, M. H., & Mohammadi, B. (2018). Pan evaporation prediction using a hybrid multilayer perceptron-firefly algorithm (MLP-FFA) model: Case study in North Iran. *Theoretical and Applied Climatology*, *133*(3–4), 1119–1131. https://doi.org/10.1007/s00704-017-2244-0

Granata, F., Papirio, S., Esposito, G., Gargano, R., & de Marinis, G. (2017). Machine learning algorithms for the forecasting of wastewater quality indicators. *Water (Switzerland)*, *9*(2), 1–12. https://doi.org/10.3390/w9020105

Hayder, G., & Mustafa, H. (2021). Cultivation of aquatic plants for biofiltration of wastewater. *Letters in Applied NanoBioScience*, *10*(1), 1919–1924. https://doi.org/10.33263/LIANBS101.19191924

Hendriarianti, E., & Soetedjo, A. (2021). IoT based real-time monitoring of phytoremediation of wastewater using the mathematical model implemented on the embedded systems. *International Journal of Intelligent Engineering and Systems*, *14*(2), 285–294. https://doi.org/10.22266/ijies2021.0430.25

Jiang, J., Tang, S., Han, D., Fu, G., Solomatine, D., & Zheng, Y. (2020). A comprehensive review on the design and optimization of surface water quality monitoring networks. *Environmental Modelling and Software*, *132*, 104792. https://doi.org/10.1016/j.envsoft.2020.104792

Kitt, F.-P., Will, P., & Robert, E. (2005). *Arizona Watershed Stewardship Guide: Water Quality & Monitoring*. College of Agriculture and Life Sciences, University of Arizona (Tucson, AZ), 18.

Koutroumanidis, T., Ioannou, K., & Arabatzis, G. (2009). Predicting fuelwood prices in Greece with the use of ARIMA models, artificial neural networks and a hybrid ARIMA-ANN model. *Energy Policy, 37*(9), 3627–3634. https://doi.org/10.1016/j.enpol.2009.04.024

Lee, J., Kim, C. G., Lee, J. E., Kim, N. W., & Kim, H. (2018). Application of artificial neural networks to rainfall forecasting in the Geum River Basin, Korea. *Water (Switzerland), 10*(1448), 1–14. https://doi.org/10.3390/w10101448

Liu, S., Tai, H., Ding, Q., Li, D., Xu, L., & Wei, Y. (2013). A hybrid approach of support vector regression with genetic algorithm optimization for aquaculture water quality prediction. *Mathematical and Computer Modelling, 58*(3–4), 458–465. https://doi.org/10.1016/j. mcm.2011.11.021

Moosavi, V., Vafakhah, M., Shirmohammadi, B., & Ranjbar, M. (2012). Optimization of wavelet-ANFIS and wavelet-ANN hybrid models by taguchi method for groundwater level forecasting. *Arabian Journal for Science and Engineering, 39*(3), 1785–1796. https:// doi.org/10.1007/s13369-013-0762-3

Mustafa, H. M., & Hayder, G. (2020a). Recent studies on applications of aquatic weed plants in phytoremediation of wastewater: A review article. *Ain Shams Engineering Journal.* https://doi.org/10.1016/j.asej.2020.05.009

Mustafa, H. M., & Hayder, G. (2020b). Performance of *Pistia stratiotes, Salvinia molesta,* and *Eichhornia crassipes* aquatic plants in the tertiary treatment of domestic wastewater with varying retention times. *Applied Sciences (Switzerland), 10*(9105), 1–19. https:// doi.org/10.3390/app10249105

Mustafa, H. M., & Hayder, G. (2021a). Cultivation of S. molesta plants for phytoremediation of secondary treated domestic wastewater. *Ain Shams Engineering Journal.* https://doi. org/10.1016/j.asej.2020.11.028

Mustafa, H. M., & Hayder, G. (2021b). Evaluation of water lettuce, giant salvinia and water hyacinth systems in phytoremediation of domestic wastewater. *H2Open Journal, 4*(1), 167–181. https://doi.org/10.2166/h2oj.2021.096

Mustafa, H. M., & Hayder, G. (2021c). Performance of Salvinia molesta plants in tertiary treatment of domestic wastewater. *Heliyon, 7*(October 2020), e06040. https://doi. org/10.1016/j.heliyon.2021.e06040

Mustafa, H. M., & Hayder, G. (2021d). Potentials in bioremediation of domestic wastewater comparison of pistia stratiotes and lemna minor plants. *Journal – The Institution of Engineers, Malaysia, 82*(1), 17–22.

Mustafa, H. M., Hayder, G., & Jagaba, A. (2021a). Microalgae: A sustainable renewable source for phytoremediation of wastewater and feedstock supply for biofuel generation. *Biointerface Research in Applied Chemistry, 11*(1), 7431–7444. https://doi. org/10.33263/BRIAC111.74317444

Mustafa, H. M., Hayder, G., Solihin, M., & Saeed, R.. (2021b). Applications of constructed wetlands and hydroponic systems in phytoremediation of wastewater. *IOP Conference Series: Earth and Environmental Science, 708*(012087). https://doi. org/10.1088/1755-1315/708/1/012087

Mustafa, H. M., Mustapha, A., Hayder, G., & Salisu, A. (2021c). Applications of IoT and Artificial Intelligence in Water Quality Monitoring and Prediction: A Review. *Proceedings of the 6th International Conference on Inventive Computation Technologies, ICICT 2021, January,* 968–975. https://doi.org/10.1109/ICICT50816.2021.9358675

Nemati, S., Fazelifard, M. H., Terzi, Ö., & Ghorbani, M. A. (2015). Estimation of dissolved oxygen using data-driven techniques in the Tai Po River, Hong Kong. *Environmental Earth Sciences, 74*(5), 4065–4073. https://doi.org/10.1007/s12665-015-4450-3

Nourani, V., Elkiran, G., & Abba, S. I. (2018). Wastewater treatment plant performance analysis using artificial intelligence - An ensemble approach. *Water Science and Technology, 78*(10), 2064–2076. https://doi.org/10.2166/wst.2018.477

Pomeroy, J. W. (2007). The CRHM model: A platform for basin proees representation and model sturcuture on physical evidence. *Hydrological Processes*, *21*(March), 4142–4152. https://doi.org/10.1002/hyp

Rajaee, T., Ahmad, S., Zounemat-Kermani, M., & Nourani, V. (2009). Daily suspended sediment concentration simulation using ANN and neuro-fuzzy models. *Science of the Total Environment*, *407*(17), 4916–4927. https://doi.org/10.1016/j.scitotenv.2009.05.016

Saravanan, K., Anusuya, E., Kumar, R., & Son, L. H. (2018). Real-time water quality monitoring using Internet of Things in SCADA. *Environmental Monitoring and Assessment*, *190*(9). https://doi.org/10.1007/s10661-018-6914-x

Taylor, K. E. (2001). Summarizing multiple aspects of model performance in a single diagram. *Journal of Geophysical Research*, *106*(D7), 7183–7192. https://doi.org/10.1029/2000JD900719

Whitehead, P., Dolk, M., Peters, R., & Leckie, H. (2019). Water Quality Modelling, Monitoring, and Management. In S. J. Dadson, D. E. Garrick, E. C. Penning-Rowsell, J. W. Hall, R. Hope, & J. Hughes (Eds.), *Water Science, Policy, and Management* (pp. 55–73). Wiley online library, Hoboken, NJ. https://doi.org/10.1002/9781119520627.ch4

Yaseen, Z. M., Ghareb, M. I., Ebtehaj, I., Bonakdari, H., Siddique, R., Heddam, S., Yusif, A. A., & Deo, R. (2018). Rainfall pattern forecasting using novel hybrid intelligent model based ANFIS-FFA. *Water Resources Management*, *32*(1), 105–122. https://doi.org/10.1007/s11269-017-1797-0

Zhu, S., & Heddam, S. (2020). Prediction of dissolved oxygen in urban rivers at the three Gorges reservoir, China: Extreme learning machines (ELM) versus artificial neural network (ANN). *Water Quality Research Journal of Canada*, *55*(1), 106–118. https://doi.org/10.2166/WQRJ.2019.053

10 Case Study

Emerging Black Box System Identification Model with Neuro-Boasting Machine Learning Techniques for Experimental Validation of Phytoremediation of Wastewater: A Data Intelligent Approach

10.1 INTRODUCTION

Water quality modelling is a complex process that attempts to quantify the dynamic interactions between lakes, river and groundwater hydrology in order to gain a better understanding of the underlying science (Whitehead et al., 2019). However, target 6.3 of the UN's Sustainable Development Goals (SDGs) is aimed at enhancing water quality through lowering pollution, preventing and reducing the release of harmful chemicals (United Nations, 2018). Thus, effective decision making and policy systems are required to address these issues on a national and worldwide scale (Whitehead et al., 2019). Wireless sensors are often used to monitor water quality parameters including oxidation–reduction potential (ORP), chlorine, pH, temperature and turbidity. In most water quality detection sensors, the sensing component is constructed based on chemical, physical and biological parameters of water (Dong et al., 2015).

Furthermore, the applications of aquatic plants (macrophytes) in phytoremediation of domestic wastewater have been reported (Hayder & Mustafa, 2021; Mustafa et al., 2021a, 2021b; Mustafa & Hayder, 2020a, 2020b, 2021b, 2021c, 2021d). Mustafa and Hayder (2021a) investigated the potentials of varying density of giant salvinia plants in tertiary treatment of wastewater. Parveen et al. (2019) developed SVM model to forecast Ni(II) ions from tea industry waste in terms of pH, flow rate, Ni(II) ion, effluent volume, contact time, particle size adsorbent and bed depth. Additionally, the authors used statistical parameters to compare the SVM and MLR models. The results obtained showed that the correlation coefficient (R) and average absolute relative error (AARE) values for the SVM-based model are 0.993 and 6.88%, respectively, while those of the MLR model were 0.8393 and 74.54%, respectively. Furthermore, Kumar et al. (2019) applied two-factor MLR in the prediction of heavy metal uptake by *P. stratiotes* from paper mill effluent. The findings indicated that the selected input variables helped in the development of prediction models with a high model efficiency (ME), higher linear regression (h^2), low mean average normalizing error (MANE) of 0.92–0.99, $h^2 > 0.72$ and MANE < 0.02, respectively. Despite this, few studies have examined the application of artificial intelligence (AI) models in wastewater treatment plants, as computational intelligence models are data-driven and their accuracy differs with several datasets. Additional research is required to establish the global promising models that can be utilized in wastewater treatment plants, particularly in phytoremediation of wastewater. Hence, the novelty and motivation of this study involved investigating and monitoring the capabilities of *S. molesta* plants in wastewater treatment and validating the experimental data using the potential of promising machine learning (ML) tools, artificial neural network (ANN), support vector machine (SVM), adaptive neuro-fuzzy inference system (ANFIS) and multilinear regression (MLR). In this vein, available data on the phytoremediation of domestic wastewater is largely based on empirical studies that investigated the efficiency of aquatic plants in the reduction of pollutants from domestic wastewater. Furthermore, there is no report in recent literature on the applications of Arduino IoT devices combined with MATLAB® software for predicting and modelling of ORP water quality parameter in phytoremediation of domestic wastewater using ANN, SVM, ANFIS and MLR models. Moreover, ORP is used in determining the oxidation and reduction potential of water samples. It is an important parameter that governs the biochemical processes in water quality monitoring. ORP is crucial in water quality monitoring because it indicates the level of pollutants present in water based on the oxidation and reduction properties such as microbial disinfection, nitrification and denitrification reactions, manganese and iron compound precipitation, and corrosion of drinking water distribution pipes (Mamun et al., 2019). A positive ORP value indicates that the substance is an oxidizing agent, whereas a negative reading suggests that it is a reducing agent.

10.2   MATERIALS AND METHODOLOGY

10.2.1   Water Quality Parameters Used

The water quality parameters used in the modelling and evaluation of the turbidity (TURBt) are presented in Table 10.1.

10.2.2   Data Processing and Statistical Analysis

Data preprocessing is very crucial in any data-driven model. Different data process was examined and analysed before the model preparation. The data preprocessing

TABLE 10.1

Parameter Used (Influent (Raw) and Effluent (Treated) Samples)

Parameters	Influents Parameter	Effluent Parameter
Raw turbidity	TURBr	ORPt
Treated turbidity	TURBt	
Raw total dissolve solid	TDSr	
Treated total dissolve solid	TDSt	
Raw oxidation–reduction potential	ORPr	
Treated oxidation–reduction potential	ORPt	
Raw temperature	TEMPr	
Treated temperature	TEMPt	

and feature selection stages are critical in the initial step of developing AI models. This procedure has a considerable effect on the accuracy of prediction in any sort of data (Hayder et al., 2020; Hsu et al., 2011). The 2211 time-series data retrieved from the measurement of ORP (the influent and effluent of *S. molesta* treatment systems) for 2 weeks was used in this study. Therefore, the generated data was preprocessed using descriptive statistics, sensitivity analysis and normalization. As with other AI algorithms, ANNs benefit from data standardization. The input data is usually adjusted to standardize the scale of each variable. In this study, the data was normalized to the range of [0,1] using Equation 10.1 prior to the training (Hayder et al., 2020; Hsu et al., 2011).

$$X_{norm} = \frac{x - x_{min}}{x_{max} - x_{min}} \tag{10.1}$$

where x, X_{norm}, x_{min} and x_{max} are the observed value, normalized value, the minimum and maximum values of the variable, respectively.

In terms of feature selection, if the correlation score is less than |0.5|, it implies that there is a low correlation (weak association) between the input and the target variable. Furthermore, the validation methods were implemented using a different approach. The holdout approach was used in this study. The data was randomly assigned to two sets: calibration and validation (verification), which is a variation on *k*-fold cross-validation (Tsioptsias et al., 2016).

10.2.3 Artificial Neural Network (ANN)

In this study, feedforward neural network (FFNN) structures were implemented, and their effectiveness was evaluated and compared with the SVM, ANFIS and multilinear model (MLR). The proposed two-model combinations (M1 and M2) were trained with three-layer (input, hidden and output) ANN architecture in which the number of hidden nodes is greater than the number of inputs. The FFNN was built for predicting the ORP of the effluent water samples using four input variables, X_i ($I = 1, ..., 4$) (temperature, turbidity, TDS and ORP of the influent samples). The $n + 1$ formula was used in selecting the hidden nodes, since it has a better degree of generalization that helps in preventing over-fitting issues. As a result, insufficient nodes can hinder the performance training and validation of the network (Ahmed & Sarma, 2007; Guo et al., 2015).

10.2.4 CONCEPT OF SUPPORT VECTOR MACHINE (SVM)

In this study, the model prediction was generated using SVM. Additionally, the two models were trained using cubic, linear, fine Gaussian, polynomial, coarse Gaussian and medium Gaussian kernel with a cross-validation of 10. The kernel functions were trained with more than 100 iterations. After the training, it was discovered that the fine Gaussian kernel resulted in the optimum fitness model for predicting the effluent water quality. For a given set of training data $\{(x_i, d_i)\}_i^N$ (x_i is the input vector, d_i is the actual value and N is the total number of data patterns), the general SVM function is given in Equations 10.2–10.7.

$$y = f(x) = w\varphi(x_i) + b \qquad (10.2)$$

$$\text{Minimize} : \frac{1}{2}\|w\|^2 + C\left(\sum_i^N (\xi_i + \xi_i^*)\right) \qquad (10.3)$$

$$\text{Subject to} : \begin{cases} w_i\varphi(x_i) + b_i - d_i \leq \varepsilon + \xi_i^* \\ d_i - w_i\varphi(x_i) + b_i \leq \varepsilon + \xi_i^* \quad i = 1, 2, ..., N \\ \xi_i, \xi_i^* \end{cases} \qquad (10.4)$$

$$w^* = \sum_{i=1}^N (\alpha_i - \alpha_i^*)\varphi(x_i) \qquad (10.5)$$

So, the final form of SVM can be expressed as:

$$f(x, \alpha_i, \alpha_i^*) = \sum_{i=1}^N (\alpha_i - \alpha_i^*)K(x, x_i) + b \qquad (10.6)$$

$$k(x_1, x_2) = \exp(-\gamma\|x_1 - x_2\|^2) \qquad (10.7)$$

where γ is the kernel parameter.

10.2.5 CONCEPT OF ADAPTIVE NEURO-FUZZY INFERENCE SYSTEM (ANFIS)

ANFIS is a FFNN structure comprised of neuro-fuzzy system components at each layer. The ANFIS model utilizes the fuzzy Sugeno model within an adaptive system framework to promote learning and adaptation. The fuzzy inference applied in this study has four inputs and one output for Model 1 (M1), and three inputs and one output for Model 2 (M2). ANFIS architecture can be described by assuming that there are two inputs designated as x and y. Then the two fuzzy if–then rules for a first-order Sugeno fuzzy model can be expressed as follows:

$$\text{Rule 1: If } (x \text{ is } A_1) \text{ and } (y \text{ is } B_1), \text{ then } (f_1 = p_1 x + q_1 y + r_1), \tag{10.8}$$

$$\text{Rule 2: If} (x \text{ is } A_2) \text{ and } (y \text{ is } B_2), \text{ then } (f_2 = p_2 x + q_2 y + r_2), \tag{10.9}$$

where A_i and B_i are fuzzy sets, f_i is the output, and p_i, q_i, and r_i are the design parameters that are determined during the training process. The ANFIS architecture used to implement the two rules is shown below:

$$\text{Layer 1}: Q_i^1 = \mu_{Ai}(x) \quad \text{for } i = 1,2 \quad \text{or} \quad Q_i^1 = \mu_{Bi}(x) \quad \text{for } i = 3,4 \tag{10.10}$$

$$\text{Layer 2}: Q_i^2 = w_i = \mu_{Ai}(x) \quad \mu_{Bi}(y) \quad \text{for } i = 1,2 \tag{10.11}$$

$$\text{Layer 3}: Q_i^3 = \bar{w}_i = \frac{w_i}{w_1 + w_2} \quad i = 1,2 \tag{10.12}$$

$$\text{Layer 4}: Q_i^4 = \bar{w}_i(p_i x + q_i y + r_i) = \bar{w}_i f_i \tag{10.13}$$

p_1, q_1, and r_1 are irregular parameters called consequent parameters.

$$\text{Layer 5}: Q_i^5 = \bar{w}_i(p_i x + q_i y + r_i) = \sum_i \bar{w}_i f_i = \frac{\sum_i w_i f_i}{\sum_i w_i} \tag{10.14}$$

For the purpose of this study, the optimum hyperparameter tuning for developing the ANFIS model especially the membership functions (MF) (Gaussian, trapezoidal, sigmoid and triangular etc.) was selected using trial-and-error techniques. The choice of the best training algorithms between backpropagation and hybrid learning algorithms was both employed for optimizing the parameters of the MFs; subsequently, hybrid emerged as the merit training algorithm. The error tuning was set to be 0.005 and the iterations were between the range of 30 and 70 for both the training and testing phase.

10.2.6 MULTILINEAR REGRESSION (MLR)

MLR is used in determining the level of variation between two or more variables. MLR of Y (dependent variable) on X (independent variable) is defined as (Chen & Liu, 2015):

$$y = b_0 + b_1 x_1 + b_2 x_2 + b_3 x_3 + \cdots + b_i x_i \tag{10.15}$$

where x_1 is the value of the ith predictor, b_0 is the regression constant and b_i is the coefficient of the ith predictor.

10.3 MODEL VALIDATION AND PERFORMANCE EVALUATION

The model performance was determined by splitting the predicted data obtained from the ANN, SVM, ANFIS and MLR simulations into 70% training (1548) and 30% testing (663) data. To monitor the performance of the models, different evaluation criteria namely coefficient of correlation (R), coefficient of determination (R^2), mean square error (MSE) and root mean square error (RMSE) were used. The model performance parameters are formulated as follows:

$$R = \sqrt{1 - \frac{\sum_{i=1}^{n}(A_i - P_i)^2}{\sum_{i=1}^{n}(A_i - \bar{A})^2}} \tag{10.16}$$

$$R^2 = 1 - \frac{\sum_{i=1}^{n}(A_i - P_i)^2}{\sum_{i=1}^{n}(A_i - \bar{A})^2} \tag{10.17}$$

$$MSE = \frac{1}{n}\sum_{i=1}^{n}(A_i - P_i)^2 \tag{10.18}$$

$$RMSE = \sqrt{\frac{1}{n}\sum_{i=1}^{n}(A_i - P_i)^2} \tag{10.19}$$

where A_i, $\bar{A}$ and P_i represent the actual values in ith daily, the average of actual values and predicted values, respectively, while n represents the number of data.

10.4 RESULTS AND DISCUSSION

This section discusses the outcome of the prediction and evaluation for the ORP water quality parameters of the *S. molesta* treatment system.

10.4.1 Descriptive Statistical Analysis

In this research work, the estimation of the chosen parameters was made up of different sets of input and output variables as shown in the statistical Table 10.2. In addition, the degree and type of relationship between the influent and effluent samples was investigated using the correlation matrix and descriptive statistical analysis. R was evaluated to determine the extent of linear interaction between the influent and effluent variables. The concept of R is to predict a variable using a linear function that acts as a tentative indicator of a possible correlation between the set of variables. As presented in Table 10.3, the R-value in bold shows the stationary and significant variables with a probability less than 0.05 ($p < 0.05$). This indicates a high degree of linear correlation. Additionally, the R values with the negative sign suggests that the two variables have an inverse interaction between them. Therefore, the weakness observed by R-value showed that it is important to apply more robust methods in complex modelling interactions. Furthermore, the evaluation of R was also important in the pre-analysis of the data and generation of the data-driven models, since the directional sign (positive or negative) shows the relationship between the dimension and proportionality of the variables (dependent and independent).

As mentioned earlier, two different models based on dominant factors were generated to predict the ORPt of the *S. molesta* treatment system. This could be better understood by considering the correlation matrix between the parameters as indicated in Figure 10.1.

The choice of two input combinations is primarily focused on the analysis of linear sensitivity techniques. Several authors have reported the significance of applying the non-linear sensitivity input variables selection methods in determining the most critical factors (Hadi et al., 2019; Nourani & Sayyah, 2012).

From Figure 10.1, the Spearman Pearson Correlation demonstrates how effectively a linear function can be used to explain the relationship between the water quality variables. The direction or the sign does not affect the strength of the correlation. In other words, a positive correlation implies that an increase in the first parameter is proportional to an increase in the second parameter, whereas a negative coefficient implies an inverse relationship, in which as one parameter increases, the second parameter decreases (Abba et al., 2020a; Ghali et al., 2020; Eisinga et al., 2013).

Except for ORPr, and TEMPr in the case of TDSt and TURBt, respectively, all the WQ variables displayed satisfactory inverse relationship with the target variables, although previous studies (Elkiran et al., 2019; Hadi et al., 2019) had discouraged the application of classical linear input variable selection methods and suggested the use of non-linear methods. The non-linear methods can be used for selection of inputs and determination of linear patterns between two or more variables. Additionally, according to Figure 10.1, TEMPr (0.5) has the highest correlation coefficient with TDSt, while ORPr emerged as the least correlated variable. In the case of TURBr, TEMPt has the highest positive relationship with the same TEMPr emerged lowest. For ORPt, the highest positive variables were ORPr. This kind of relationship is quite non-linear in nature and, hence, requires urgent attention of non-linear models to address them. This statement is in line with various studies in the field of WWTP (Abba & Elkiran, 2017).

TABLE 10.2
Descriptive Statistic of the Parameters

	TURBr	TURBt	TDSt	TDSr	ORPr	ORPt	TEMPr	TEMPt
Mean	4.6522	3.7598	79.0238	35.5773	572.6749	285.6227	44.1265	43.5283
Standard error	0.0054	0.0164	0.6284	0.3589	3.9663	3.5014	0.1406	0.1499
Median	4.7600	3.7400	75.5700	29.6300	584.8800	211.3500	43.5000	42.3200
SD	0.2541	0.7715	29.5489	16.8745	186.5008	164.6383	6.6096	7.0489
Sample variance	0.0646	0.5952	873.1404	284.7494	34782.5398	27105.7829	43.6874	49.6866
Kurtosis	4.3794	-1.0926	-0.7011	-0.2808	7.6489	1.3723	-1.0264	-1.0320
Skewness	-2.2656	-0.2762	0.0169	0.7754	0.5966	1.3272	0.1182	0.0811
Minimum	3.4700	2.0000	0.0000	1.4700	20.2800	26.5700	28.3400	28.5000
Maximum	4.9600	4.7700	134.8500	102.9300	2000.0000	1152.8300	55.3300	55.1800

TABLE 10.3
Correlation Analysis between the Input and Output Variables

Variables	TEMPr	TURBr	TDSr	ORPr	TEMPt	TURBt	TDSt	ORPt
TEMPr	1							
TURBr	0.430423	1						
TDSr	-0.79781	-0.30977	1					
ORPr	0.221699	0.471303	-0.27129	1				
TEMPt	-0.027	-0.39745	0.17753	-0.28988	1			
TURBt	0.042232	0.234459	-0.27136	0.177549	-0.59829	1		
TDSt	0.513949	0.172227	-0.40571	-0.01947	0.07073	-0.17134	1	
ORPt	-0.23246	0.269886	0.189318	0.315219	-0.33457	0.289099	-0.46082	1

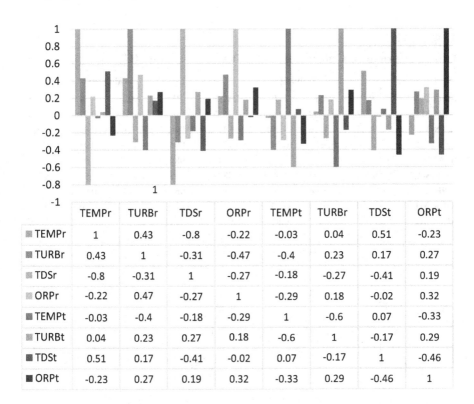

	TEMPr	TURBr	TDSr	ORPr	TEMPt	TURBr	TDSt	ORPt
TEMPr	1	0.43	-0.8	-0.22	-0.03	0.04	0.51	-0.23
TURBr	0.43	1	-0.31	-0.47	-0.4	0.23	0.17	0.27
TDSr	-0.8	-0.31	1	-0.27	-0.18	-0.27	-0.41	0.19
ORPr	-0.22	0.47	-0.27	1	-0.29	0.18	-0.02	0.32
TEMPt	-0.03	-0.4	-0.18	-0.29	1	-0.6	0.07	-0.33
TURBt	0.04	0.23	0.27	0.18	-0.6	1	-0.17	0.29
TDSt	0.51	0.17	-0.41	-0.02	0.07	-0.17	1	-0.46
ORPt	-0.23	0.27	0.19	0.32	-0.33	0.29	-0.46	1

FIGURE 10.1 Visualized correlation matrix between the parameters.

10.4.2 RESULTS OF ORPT (ANN, SVM, ANFIS AND MLR)

The performance efficiency of the ML models was obtained by comparing the observed and predicted data. In this study, different evaluation criteria were used to assess the ORP of the influent and effluent samples from the *S. molesta* treatment system. The ORPt performance criteria for ANN, SVM, ANFIS and MLR are presented in Table 10.4.

From Table 10.4, it could be observed that SVM-M1 was the best with the lowest MSE criteria. The overall outcomes proved merit for all the ML models. The average accuracy of ML models over the linear models was a 7% increment. This outcome shows the weakness of MLR, contrary to the promising ability of linear models reported in different literature.

Furthermore, Figures 10.2 and 10.3 display the ORP time-series plot for ANN, ANFIS, SVM and MLR for M1 and M2 at the validation phase. The time-series plot showed that the patterns of the observed and predicted values for M1 and M2 were found to be close for the AI models, while in the case of the MLR time-series plot, a disparity was observed in the trends of the observed and predicted data for both M1 and M2 models. Hence, this outcome corroborates with the findings presented in Table 10.4. Additionally, the developed models can play a significant and pivotal

TABLE 10.4

ORPt Performance Criteria for ANN, SVM, ANFIS and MLR

		Training				Validation		
Models	R^2	MSE	R	RMSE	R^2	MSE	R	RMSE
ANN-M1	0.9972	0.0652	0.9986	0.2553	0.9983	0.0648	0.9992	0.2545
ANN-M2	0.9825	0.4116	0.9912	0.6416	0.9803	1.7860	0.9901	1.3364
SVM-M1*	0.9990	0.0231	0.9995	0.1521	0.9988	0.1045	0.9994	0.3233
SVM-M2	0.9904	0.2254	0.9952	0.4747	0.9864	1.2346	0.9932	1.1111
ANFIS-M1	0.9898	0.2394	0.9949	0.4893	0.9939	0.5498	0.9970	0.7415
ANFIS-M2	0.9373	1.4764	0.9681	1.2151	0.9619	3.4473	0.9808	1.8567
MLR-M1	0.8710	3.0341	0.9332	1.7419	0.9221	7.0517	0.9603	2.6555
MLR-M2	0.8371	3.8357	0.9149	1.9585	0.9010	8.9557	0.9492	2.9926

* Signifies the best model.

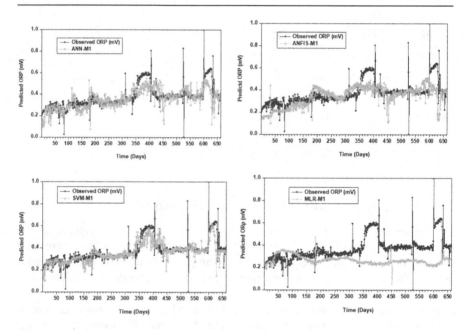

FIGURE 10.2 ORP time-series plot for ANN-M1, ANFIS-M1, SVM-M1 and MLR-M1 for validation phase.

role in providing an inexpensive and fast method of WQ monitoring compared to laboratory analysis. These ML models are robust and can predict water pollution and serve as a guide for authorized government and non-governmental agencies to develop effective strategies and manipulations for better water sustainability and management.

Furthermore, the prediction accuracy of ANN, SVM, ANFIS and MLR models for ORP was illustrated using a radar chart in Figure 10.4.

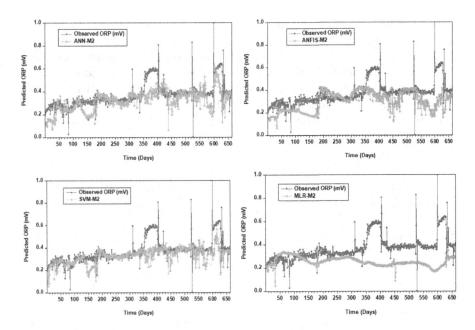

FIGURE 10.3 ORP time-series plot for ANN-M2, ANFIS-M2, SVM-M2 and MLR-M2 for validation phase.

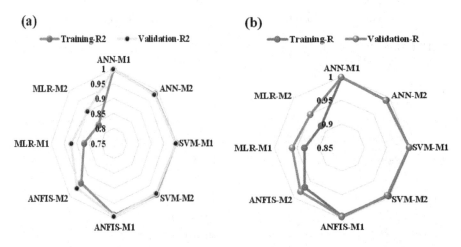

FIGURE 10.4 Radar chart for (a) goodness-of-fit and (b) correlation coefficient for ORP.

From Figure 10.4, the lowest and highest value of *R* from the models in the validation phase was found to be 0.8371 and 0.9990, respectively. The radar charts strengthen the justification of the performance criteria outcome presented in Table 10.4. Abba et al. (2020b) stated that the high value of *R* plays a significant role in selecting the best performing model.

Additionally, the Taylor diagram was used for diagnostic analysis of the models. Taylor diagram provided the visual assessment of the model's performance by illustrating three distinct scenarios (R, RMSE and normalized standard deviation). Figure 10.5 represents the Taylor diagrams for ORP for the training and validation phase. The Taylor diagram further proved the superiority of SVM-M1 in the ORP model over ANN, SVM-M2, ANFIS and MLR models.

Additionally, Figure 10.6 represents the error plot of the performance indicator for the training and validation phase.

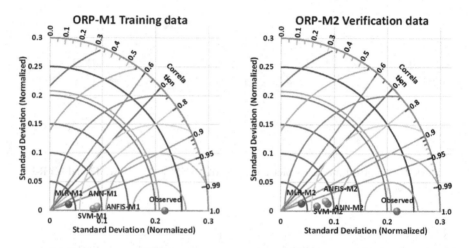

FIGURE 10.5 Two-dimensional Taylor diagram between the observed and predicted ORP in the training and verification phase.

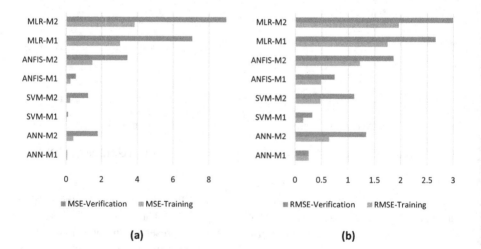

FIGURE 10.6 Error plot in both training and validation for (a) MSE and (b) RMSE.

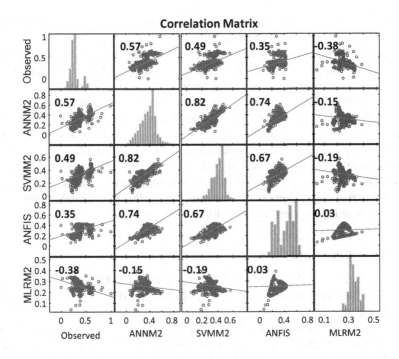

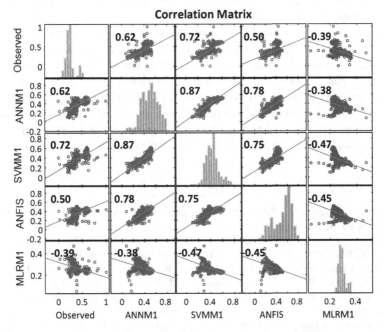

FIGURE 10.7 The marginal internal correlation plot between the observed and predicted models in the validation phase.

It was observed that the simulated ORPt achieved better fit using an ML model with the ranking order of SVM > ANN > ANFIS > MLR for both M1 and M2. Additionally, the mathematical comparison of ML models with regard to RMSE depicted that SVM-M1 increases the prediction accuracy of ANFIS-M1 and ANN-M1 by 5%. Thus, the non-linear black box revealed an accurate and valuable forecasting capability in the ORPt model. Moreover, the overall poor performance exhibited by the MLR for M1 and M2 models can be attributed to the fact that the MLR techniques are based on the assumptions of linear input–output relationship. A similar result was reported by Gaya et al. (2020). Chen and Liu (2015) applied ANN, ANFIS and MLR models to estimate DO concentrate in the Feitsui Reservoir of Northern Taiwan. The result shows that MLR model was not able to estimate the DO. The marginal internal correlation plot between the observed and predicted models in the validation phase is presented in Figure 10.7.

10.5 CONCLUSION

This research applied ML tools (ANN, SVM, ANFIS) and MLR in modelling and predicting of the ORP parameters of the *S. molesta* treated wastewater. It was found that both the classical and AI-based models were capable of modelling the ORPt. This was proved by considering the statistical indicators (R, R^2, MSE and RMSE). The visual investigation of the models indicated that model 1 (M1) is superior to M2 for all the models. This is attributed to the fact that M1 contained additional input variables which served as the dominant and significant variables. Additionally, it could be observed that SVM-M1 performed best with the lowest MSE criteria. The overall outcomes proved merit for all the AI-based models. The average accuracy of the ML models over the linear models was a 7% increment. Similarly, the developed models can play a significant and pivotal role in providing an inexpensive and fast method of water quality monitoring compared to the traditional laboratory analysis. These AI models are robust and can predict water pollution, and serve as a guide for authorized government and non-governmental agencies to develop effective strategies and manipulations for better water sustainability and management.

REFERENCES

Abba, S. I., & Elkiran, G. (2017). Effluent prediction of chemical oxygen demand from the astewater treatment plant using artificial neural network application. *Procedia Computer Science, 120*, 156–163. https://doi.org/10.1016/j.procs.2017.11.223

Abba, S. I., Elkiran, G., & Nourani, V. (2020a). Non-Linear Ensemble Modeling for Multi-Step Ahead Prediction of Treated COD in Wastewater Treatment Plant. In R. Aliev, J. Kacprzyk, W. Pedrycz, M. Jamshidi, M. Babanli, & F. Sadikoglu (Eds.), *10th International Conference on Theory and Application of Soft Computing, Computing with Words and Perceptions - ICSCCW-2019. ICSCCW 2019. Advances in Intelligent Systems and Computing* (Vol. 1095, Issue November 2019, pp. 683–689). Springer, Cham. https://doi.org/10.1007/978-3-030-35249-3_88

Abba, S. I., Linh, N. T. T., Abdullahi, J., Ali, S. I. A., Pham, Q. B., Abdulkadir, R. A., Costache, R., Van Nam, T., & Anh, D. T. (2020b). Hybrid machine learning ensemble techniques for modeling dissolved oxygen concentration. *IEEE Access, 8*, 157218– 157237. https://doi.org/10.1109/access.2020.3017743

Ahmed, J. A., & Sarma, A. K. (2007). Artificial neural network model for synthetic streamflow generation. *Water Resources Management, 21*(6), 1015–1029. https://doi.org/10.1007/s11269-006-9070-y

Chen, W., & Liu, W. (2015). Water quality modeling in reservoirs using multivariate linear regression and two neural network models. *Advances in Artificial Neural Systems,* 2015(521721), 1–12. https://doi.org/10.1155/2015/521721 Research

Dong, J., Wang, G., Yan, H., Xu, J., & Zhang, X. (2015). A survey of smart water quality monitoring system. *Environmental Science and Pollution Research, 22*(7), 4893–4906. https://doi.org/10.1007/s11356-014-4026-x

Eisinga, R., Te Grotenhuis, M., & Pelzer, B. (2013). The reliability of a two-item scale: Pearson, Cronbach, or Spearman-Brown? *International Journal of Public Health, 58*(4), 637–642. https://doi.org/10.1007/s00038-012-0416-3

Elkiran, G., Nourani, V., & Abba, S. I. (2019). Multi-step ahead modelling of river water quality parameters using ensemble artificial intelligence-based approach. *Journal of Hydrology, 577*(April), 1–12. https://doi.org/10.1016/j.jhydrol.2019.123962

Gaya, M. S., Abba, S. I., Abdu, A. M., Tukur, A. I., Saleh, M. A., Esmaili, P., & Wahab, N. A. (2020). Estimation of water quality index using artificial intelligence approaches and multi-linear regression. *IAES International Journal of Artificial Intelligence, 9*(1), 126–134. https://doi.org/10.11591/ijai.v9.i1.pp126-134

Ghali, U. M., Usman, A. G., Chellube, Z. M., Degm, M. A. A., Hoti, K., Umar, H., & Abba, S. I. (2020). Advanced chromatographic technique for performance simulation of anti-Alzheimer agent: An ensemble machine learning approach. *SN Applied Sciences, 2*(1871), 1–12. https://doi.org/10.1007/s42452-020-03690-2

Guo, H., Jeong, K., Lim, J., Jo, J., Kim, Y. M., Park, J.-P., Kim, J. H., & Cho, K. H. (2015). Prediction of effluent concentration in a wastewater treatment plant using machine learning models. *Journal of Environmental Sciences (China), 32*(April), 90–101. https://doi.org/10.1016/j.jes.2015.01.007

Hadi, S. J., Abba, S. I., Sammen, S. S., Salih, S. Q., Al-Ansari, N., & Yaseen, Z. M. (2019). Non-linear input variable selection approach integrated with non-tuned data intelligence model for streamflow pattern simulation. *IEEE Access, 7,* 141533–141548. https://doi.org/10.1109/ACCESS.2019.2943515

Hayder, G., & Mustafa, H. (2021). Cultivation of aquatic plants for biofiltration of wastewater. *Letters in Applied NanoBioScience, 10*(1), 1919–1924. https://doi.org/10.33263/LIANBS101.19191924

Hayder, G., Solihin, M., & Mustafa, H. (2020). Modelling of river flow using particle swarm optimized cascade-forward neural networks: A case study of Kelantan river in Malaysia. *Applied Sciences (Switzerland), 10*(23), 1–16. https://doi.org/10.3390/app10238670

Hsu, H. H., Hsieh, C. W., & Da Lu, M. (2011). Hybrid feature selection by combining filters and wrappers. *Expert Systems with Applications, 38*(7), 8144–8150. https://doi.org/10.1016/j.eswa.2010.12.156

Kumar, V., Singh, J., & Kumar, P. (2019). Heavy metal uptake by water lettuce (Pistia stratiotes L.) from paper mill effluent (PME): Experimental and prediction modeling studies. *Environmental Science and Pollution Research, 26,* 14400–14413.

Mamun, K. A., Islam, F. R., Haque, R., Khan, M. G. M., Prasad, A. N., & Haqva, H. (2019). Smart water quality monitoring system design and KPIs analysis: Case sites of fiji surface water. *Sustainability (Switzerland), 11*(7110), 1–21.

Mustafa, H. M., & Hayder, G. (2020a). Performance of Pistia stratiotes, Salvinia molesta, and Eichhornia crassipes aquatic plants in the tertiary treatment of domestic wastewater with varying retention times. *Applied Sciences (Switzerland), 10*(9105), 1–19. https://doi.org/10.3390/app10249105

Mustafa, H. M., & Hayder, G. (2020b). Recent studies on applications of aquatic weed plants in phytoremediation of wastewater: A review article. *Ain Shams Engineering Journal.* https://doi.org/10.1016/j.asej.2020.05.009

Mustafa, H. M., & Hayder, G. (2021a). Cultivation of S. molesta plants for phytoremediation of secondary treated domestic wastewater. *Ain Shams Engineering Journal, 12,* 2585–2592. https://doi.org/10.1016/j.asej.2020.11.028

Mustafa, H. M., & Hayder, G. (2021b). Evaluation of water lettuce, giant salvinia and water hyacinth systems in phytoremediation of domestic wastewater. *H2Open Journal, 4*(1), 167–181. https://doi.org/10.2166/h2oj.2021.096

Mustafa, H. M., & Hayder, G. (2021c). Performance of Salvinia molesta plants in tertiary treatment of domestic wastewater. *Heliyon, 7*(October 2020), e06040. https://doi.org/10.1016/j.heliyon.2021.e06040

Mustafa, H. M., & Hayder, G. (2021d). Comparison of *Pistia stratiotes* and *Lemna minor* plants potentials in bioremediation of domestic wastewater. *Journal – The Institution of Engineers, Malaysia, 82*(1), 17–22.

Mustafa, H. M., Hayder, G., & Jagaba, A. (2021a). Microalgae: A sustainable renewable source for phytoremediation of wastewater and feedstock supply for biofuel generation. *Biointerface Research in Applied Chemistry, 11*(1), 7431–7444. https://doi.org/10.33263/BRIAC111.74317444

Mustafa, H. M., Hayder, G., Solihin, M., & Saeed, R. (2021b). Applications of constructed wetlands and hydroponic systems in phytoremediation of wastewater. *IOP Conference Series: Earth and Environmental Science, 708*(012087). https://doi.org/10.1088/1755-1315/708/1/012087

Nourani, V., & Sayyah, M. (2012). Sensitivity analysis of the artificial neural network outputs in simulation of the evaporation process at different climatologic regimes. *Advances in Engineering Software, 47,* 127–129. https://doi.org/10.1016/j.advengsoft.2011.12.014

Parveen, N., Zaidi, S., & Danish, M. (2019). Support vector regression (SVR)-based adsorption model for Ni (II) ions removal. *Groundwater for Sustainable Development, 9,* 100232. https://doi.org/10.1016/j.gsd.2019.100232

Tsioptsias, N., Tako, A., & Robinson, S. (2016). Model Validation and Testing in Simulation: A Literature Review. In B. Hardy, A. Qazi, & S. Ravizza (Eds.), *5th Student Conference on Operational Research (SCOR'16). Open Access Series in Informatics* (Vol. 50, Issue 6, pp. 6.1–6.11). Schloss Dagstuhl – Leibniz-Zentrum für Informatik, Dagstuhl Publishing, Wadern, Germany. https://doi.org/10.4230/OASIcs.SCOR.2016.6

United Nations. (2018). *The 2030 Agenda and the Sustainable Development Goals an Opportunity for Latin America and the Caribbean (LC/G.2681-P/Rev.3), Santiago.* https://repositorio.cepal.org/bitstream/handle/11362/40156/25/S1801140_en.pdf

Whitehead, P., Dolk, M., Peters, R., & Leckie, H. (2019). Water Quality Modelling, Monitoring, and Management. In S. J. Dadson, D. E. Garrick, E. C. Penning-Rowsell, J. W. Hall, R. Hope, & J. Hughes (Eds.), *Water Science, Policy, and Management* (pp. 55–73). https://doi.org/10.1002/9781119520627.ch4

Index

Note: **Bold** page numbers refer to tables and *italic* page numbers refer to figures.

Printed in the United States
by Baker & Taylor Publisher Services